The Defensible Evidence Framework™

A Practical Model for Demonstrating Reasonable Cybersecurity Governance

Victor M. Font Jr.

Copyright

First Edition

2026

Published by

A FontLife Publication, LLC

Raleigh, North Carolina, USA

ISBN

Paperback: 978-1-62422-071-5

eBook: 978-1-62422-072-2

Hardback: 978-1-62422-073-9

Disclaimer

This publication is intended solely for educational and informational purposes. It does not constitute legal, regulatory, financial, accounting, cybersecurity, or other professional advice.

Although every reasonable effort has been made to ensure the accuracy of the information presented, cybersecurity threats, regulations, standards, and legal interpretations continually evolve. Readers should consult qualified legal, regulatory, audit, insurance, or cybersecurity professionals regarding their specific circumstances.

Neither the author nor the publisher assumes any responsibility or liability for any loss,

Printed in the United States of America

Published by A FontLife Publication, LLC

https://cybergovernancecenter.com

Contents

Foreword

There was a time when cybersecurity could reasonably be viewed as an operational concern. Organizations built networks, deployed firewalls, purchased antivirus software, and entrusted technical specialists with protecting digital assets. Boards were informed when necessary, budgets were approved, and oversight largely ended there.

That time has passed.

Today, every organization depends on technology to fulfill its mission. Whether serving customers, patients, students, citizens, donors, or shareholders, digital systems have become inseparable from organizational success. Consequently, cybersecurity is no longer simply a technical discipline—it is a matter of governance.

The consequences of governance failure have become impossible to ignore. Data breaches erase public trust. Ransomware halts critical operations. Regulatory investigations examine not only what happened, but how leadership exercised oversight before the incident occurred. Directors are increasingly expected to demonstrate that cybersecurity risks were understood, discussed, prioritized, and governed with the same rigor applied to financial stewardship and legal compliance.

Yet many Boards remain at a disadvantage.

Foreword

Cybersecurity literature overwhelmingly focuses on technology, threat actors, frameworks, controls, and implementation. Valuable though these resources are, they often leave Directors searching for guidance that speaks to their actual responsibilities. Boards are not responsible for configuring security controls. They are responsible for exercising fiduciary oversight.

This book addresses that distinction.

Rather than teaching readers how to defend networks, it explains how leaders govern cybersecurity as an enterprise responsibility. It frames cyber risk through the lens of fiduciary duty, organizational accountability, strategic decision-making, policy authority, assurance, resilience, and evidence. It recognizes that effective governance is measured not by technical perfection, but by informed judgment, disciplined oversight, and defensible decision-making.

Throughout these pages, the reader will find a practical governance model grounded in established principles rather than fleeting technology trends. The concepts discussed here apply regardless of organizational size, sector, or industry because the responsibilities of governance remain remarkably consistent. Whether serving on the board of a public corporation, nonprofit organization, healthcare system, financial institution, educational institution, or government-adjacent entity, Directors are increasingly expected to answer the same fundamental questions:

Did leadership understand the organization's cyber risks?
Were appropriate governance structures established?
Were material risks communicated and acted upon?
Can the organization demonstrate reasonable oversight?

Those questions define modern cybersecurity governance.

Foreword

The purpose of this book is not to create cybersecurity experts. Its purpose is to help Directors become more effective governors. It provides a framework for asking better questions, making better decisions, and creating governance systems capable of withstanding regulatory scrutiny, legal examination, and public accountability.

Cybersecurity will continue to evolve. Technologies will change. Threat actors will adapt. Regulations will mature. What will not change is the Board's responsibility to govern with diligence, integrity, and foresight.

That responsibility begins here.

Preface

Every significant cybersecurity incident eventually arrives at the same destination.

It begins with technical questions.

How did the attacker get in?

Which systems were affected?

What data was compromised?

Were backups available?

Could the attack have been prevented?

These questions dominate the first hours and days of an incident. Security teams investigate, forensic specialists reconstruct events, and technical experts work to contain the damage. Their efforts are essential, but they are only the beginning of a much larger story.

As the incident unfolds, the questions change.

What did leadership know?

When did they know it?

Preface

What decisions were made?

Who accepted the risk?

Was the Board informed?

What evidence demonstrates that reasonable oversight occurred?

At that moment, the incident is no longer just a cybersecurity event. It has become a governance event.

That realization fundamentally changed how I think about cybersecurity.

Over the course of my career, I have worked with organizations across both the public and private sectors, helping leaders improve governance, strengthen risk management, and build resilient technology organizations. I have watched organizations invest millions of dollars in sophisticated security technologies while overlooking the governance structures necessary to direct, oversee, and validate those investments. I have also seen organizations with modest budgets consistently outperform expectations because leadership exercised disciplined oversight and maintained clear accountability.

The difference was rarely the technology.

The difference was governance.

Unfortunately, governance is often misunderstood. Many organizations equate governance with compliance. Others confuse it with management or administration. Some assume governance consists merely of approving policies or receiving periodic reports. In reality, governance is the system by which leadership directs the organization, exercises oversight, evaluates performance, and demonstrates accountability.

The final point is the one that receives the least attention.

Governance must be demonstrable.

Leadership may have exercised exemplary oversight, but if that oversight cannot be demonstrated through contemporaneous evidence, regulators, insurers, auditors, investigators, courts, and other stakeholders have little basis upon which to evaluate it. Months or years after an incident, memories fade, personnel change, and informal conversations disappear. What remains is the documentary record.

That observation became the foundation for **The Defensible Evidence Framework™**.

The framework did not originate as an academic exercise. It emerged from a practical question that every governing body should ask:

If we were asked tomorrow to demonstrate that we exercised reasonable cybersecurity governance, what evidence could we produce?

The answer to that question reveals far more than an organization's security posture. It reveals the maturity of its governance.

This book expands upon the concepts first introduced in *The Defensible Evidence Framework™* white paper and develops them into a practical methodology for governing cybersecurity through evidence. While the examples throughout these pages focus on cybersecurity, the underlying principles extend beyond cyber risk. Effective governance always produces evidence. Effective leadership always leaves a defensible record of informed decision-making.

My objective is not to prescribe another compliance framework or create yet another maturity model. Countless excellent frameworks already exist. Instead, this book focuses on the evidence those frameworks should naturally produce when governance is functioning as intended.

Technology will continue to evolve. Artificial intelligence will reshape the threat landscape. Regulatory expectations will continue to expand. New frameworks will emerge while existing ones are revised. Throughout all of those changes, one principle will remain constant:

Organizations will increasingly be judged not only by what they did, but by what they can prove.

That is why this book exists.

Whether you are a Board Director, executive leader, auditor, risk professional, consultant, or cybersecurity practitioner, my hope is that these pages change the way you think about governance. Rather than asking only whether your organization is secure, I hope you begin asking whether your governance is defensible.

Because in the end, technology may prevent an incident.

Governance determines how history remembers it.

THE DEFENSIBLE EVIDENCE FRAMEWORK™

FRAMEWORK SUMMARY

Evidence-Driven Governance. Defensible Today. Resilient Tomorrow.

THE MONDAY MORNING QUESTION

"When an incident happens and the investigators arrive on Monday morning, can you prove you governed cybersecurity?"

OUR ANSWER

Yes. Because we create, collect, and preserve the right evidence every day—not just after a crisis.

THE FIVE EVIDENCE DOMAINS

1. GOVERNANCE EVIDENCE

Evidence that leadership establishes direction, structures oversight, and accepts accountability.

- Policies & Charters
- Roles & Responsibilities
- Board & Committee Oversight
- Strategic Decisions
- Governance Minutes

2. RISK EVIDENCE

Evidence of how the organization identifies, assesses, and manages cyber risk.

- Risk Assessments
- Risk Appetite & Tolerance
- Risk Registers
- Risk Treatment Decisions
- Risk Reporting

3. OVERSIGHT EVIDENCE

Evidence that leadership monitors performance and challenges assumptions.

- Executive Reporting
- Dashboards & Metrics
- Committee Reviews
- Escalations
- Challenge & Inquiry

4. OPERATIONAL EVIDENCE

Evidence that decisions are implemented through processes, controls, and activities.

- Policies & Procedures
- Control Activities
- Awareness & Training
- Third-Party Oversight
- Day-to-Day Execution

5. ASSURANCE EVIDENCE

Evidence from independent and internal assurance that governance is effective.

- Internal Audit Reports
- Independent Assessments
- Penetration Testing
- Remediation Tracking
- Continuous Monitoring

THE EVIDENCE LIFECYCLE

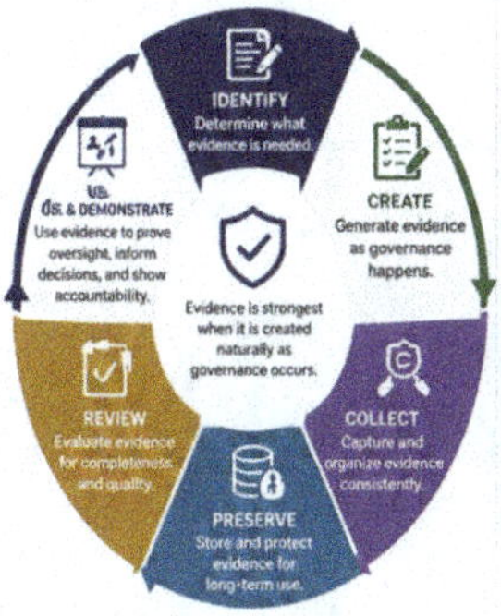

THE GOVERNANCE MATURITY LADDER

STRONGER GOVERNANCE

5 OPTIMIZED Evidence-driven culture. Continuous improvement. Anticipates future risk.

4 MANAGED Integrated governance. Risk-informed oversight. Evidence used for decisions.

3 ESTABLISHED Defined processes. Regular oversight. Evidence collected consistently.

2 DEVELOPING Basic governance in place. Evidence is ad-hoc and inconsistent.

1 INITIAL Governance is informal, reactive, and undocumented.

WEAKER GOVERNANCE

THE GOVERNANCE LOOP

A continuous cycle of evidence-driven governance.

1. **GOVERN** Establish direction, set expectations, and accept accountability.
2. **MEASURE** Monitor performance and evaluate risk exposure.
3. **IMPROVE** Act on insights and strengthen governance, controls, and resilience.
4. **PRESERVE EVIDENCE** Capture, store, and protect evidence for future use.
5. **DEMONSTRATE GOVERNANCE** Prove oversight and accountability when it matters most.

PRINCIPLES OF EVIDENCE-DRIVEN GOVERNANCE

GOVERNANCE FIRST Strong governance creates the conditions for effective cybersecurity.

EVIDENCE BY DESIGN The strongest evidence is created naturally as governance occurs—not reconstructed later.

ACCOUNTABILITY ALWAYS Clear roles, ownership, and decision rights strengthen accountability.

RISK-INFORMED Governance is based on understanding and managing risk—not eliminating it.

CONTINUOUS IMPROVEMENT Governance matures through measurement, learning, and disciplined action.

DEFENSIBLE WHEN IT COUNTS Evidence exists to demonstrate oversight and accountability under scrutiny.

OUTCOMES OF STRONG EVIDENCE-DRIVEN GOVERNANCE

- Demonstrable fiduciary duty and regulatory compliance
- Stronger incident resilience and faster recovery
- Informed decision-making and better risk management
- Greater stakeholder trust and organizational reputation
- Reduced legal, financial, and operational exposure
- Sustainable governance maturity and continuous improvement

GOOD INTENTIONS ARE NOT DEFENSIBLE.
STRONG GOVERNANCE + THE RIGHT EVIDENCE = DEFENSIBLE OVERSIGHT.

Defensible Today. Resilient Tomorrow.

Who This Book Is For

Cybersecurity is often portrayed as the responsibility of security professionals, information technology departments, or specialized technical teams. While these groups play an essential role in protecting the organization, cybersecurity itself has become far too consequential to remain solely a technical discipline.

Today, cybersecurity is a leadership responsibility.

This book is written for the people who govern organizations, make strategic decisions, allocate resources, establish priorities, and bear ultimate accountability when those decisions are questioned.

If your responsibilities include oversight, risk management, stewardship, or organizational leadership, this book is for you.

Board Directors

Members of Boards of Directors carry a fiduciary obligation to exercise informed oversight of enterprise risk. Cybersecurity has become one of the most significant enterprise risks facing organizations today, yet many Directors come from business, legal, financial, healthcare, or operational backgrounds rather than technology.

You are not expected to become a cybersecurity expert.

You are expected to ask informed questions, establish governance expectations, evaluate management's performance, and demonstrate reasonable oversight.

This book provides a governance framework for fulfilling those responsibilities with confidence.

Executive Leadership

Chief Executive Officers, Chief Operating Officers, Chief Financial Officers, General Counsel, Chief Risk Officers, Chief Information Officers, Chief Information Security Officers, and other executive leaders translate Board direction into organizational action.

The Defensible Evidence Framework™ helps executive leadership establish governance structures that are not only effective, but demonstrable. It provides a practical model for ensuring that decisions, oversight, and accountability are supported by credible evidence long before an incident occurs.

Governance, Risk, and Compliance Professionals

Professionals working in governance, enterprise risk management, internal audit, compliance, privacy, and assurance functions often serve as the bridge between organizational leadership and operational execution.

For these professionals, this book offers a structured methodology for identifying governance evidence, evaluating its completeness, and strengthening the organization's overall governance posture.

Cybersecurity Leaders

Security professionals frequently struggle to communicate cyber risk in language that resonates with executive leadership and Boards.

Rather than focusing exclusively on technical controls, this book demonstrates how cybersecurity leaders can support governance by producing evidence that enables informed oversight, strategic decision-making, and organizational accountability.

It is not another technical security manual. It is a guide for helping security leaders become trusted governance advisors.

Consultants, Auditors, and Assessors

Organizations increasingly seek independent evaluations of their cybersecurity governance programs. Whether conducting assessments, facilitating Board workshops, performing internal audits, or advising executive leadership, consultants and auditors need a practical model for evaluating governance beyond technical controls.

The Defensible Evidence Framework™ provides that model.

It offers a repeatable method for assessing whether governance activities produce sufficient evidence to demonstrate reasonable oversight.

Regulators, Insurers, and Legal Professionals

While this book is not written as a legal or regulatory text, attorneys, regulators, insurance professionals, and investigators may find value in understanding the governance principles that organizations should be implementing before incidents occur.

The framework provides a common vocabulary for discussing

governance readiness, organizational accountability, and the evidence that supports responsible leadership.

Organizations of Every Size

Although examples throughout this book frequently reference Boards and larger organizations, the principles presented here apply equally to organizations of every size.

Whether your organization has twenty employees or twenty thousand, governance still requires leadership, accountability, informed decision-making, and evidence.

The scale may differ.

The principles do not.

What This Book Assumes

This book assumes only one thing:

You believe cybersecurity deserves to be governed with the same discipline applied to finance, legal compliance, enterprise risk, and organizational strategy.

It does not assume technical expertise.

It does not assume familiarity with specific cybersecurity frameworks.

It does not require experience in information technology.

Instead, it assumes that effective governance begins with asking the right questions and creating an organizational record that demonstrates those questions were asked, considered, and acted upon.

If that describes your role—or the role you aspire to fulfill—you are the audience for this book.

Because ultimately, **The Defensible Evidence Framework™** is not about proving that an organization is perfectly secure.

It is about enabling leaders to demonstrate that they governed cybersecurity reasonably, responsibly, and with evidence.

How to Use This Book

This book is designed to be practical.

It is not intended to sit on a bookshelf as a reference consulted only after a cyber incident. Its greatest value comes from being used before an incident occurs—when leaders still have the opportunity to strengthen governance, improve oversight, and create the evidence that will ultimately demonstrate reasonable decision-making.

Whether you read this book from beginning to end or use it as a reference guide, its purpose remains the same: to help your organization build governance that is not only effective, but defensible.

Read It Sequentially

The chapters are organized intentionally.

Each section builds upon the concepts introduced in the previous one. The book begins by explaining why cybersecurity has become a governance issue before introducing the principles of the Defensible Evidence Framework™. From there, it explores each evidence domain, demonstrates how evidence is created, and concludes with practical guidance for implementation.

Readers new to cybersecurity governance will benefit most by reading the book in sequence.

Use It as a Governance Manual

Experienced Board members, executives, auditors, and consultants may find themselves returning to specific chapters as governance questions arise.

For example:

- Preparing for a Board meeting? Review the chapters on Governance Evidence and Oversight Evidence.
- Updating cybersecurity policies? Refer to the chapters discussing Governance Evidence and the Governance Evidence Repository.
- Planning an audit or assessment? Review the Assurance Evidence chapter and the Governance Readiness Scorecard.
- Preparing for a tabletop exercise or incident response review? Consult the chapters on Operational Evidence and post-incident governance.

The framework is intended to support day-to-day governance—not simply explain it.

Focus on the Questions

Throughout this book, you will encounter questions that Boards, executives, auditors, regulators, insurers, and investigators routinely ask.

Do not treat these as academic exercises.

Use them to evaluate your own organization.

If you cannot answer a question confidently—or cannot identify evidence supporting the answer—you have likely discovered an opportunity to strengthen governance.

Sometimes the most valuable outcome of reading this book is not finding answers.

It is discovering the questions that have never been asked.

Think in Terms of Evidence

One of the central themes of this book is that governance naturally produces evidence.

As you read, resist the temptation to ask:

> *"Do we have this document?"*

Instead, ask:

> *"What governance activity should naturally create this evidence?"*

That subtle shift changes the conversation.

Organizations should not create evidence simply to satisfy auditors or regulators. They should govern effectively. When governance is functioning properly, evidence becomes a natural by-product of leadership rather than an administrative burden assembled after the fact.

Adapt the Framework to Your Organization

No two organizations share identical governance structures.

Public companies answer to shareholders.

Healthcare organizations protect patients.

Financial institutions safeguard the financial system.

Government agencies serve the public.

Nonprofits steward donor resources and advance mission objectives.

The Defensible Evidence Framework™ is intentionally framework-agnostic and sector-neutral. Its principles can be adapted regardless of organizational size, mission, or regulatory environment.

The objective is not uniformity.

The objective is demonstrable governance.

Build Over Time

Few organizations will discover that every element of the framework already exists.

That is normal.

Treat the framework as a roadmap rather than a compliance checklist.

Begin by identifying evidence that already exists.

Then identify gaps.

Prioritize improvements.

Strengthen governance incrementally.

Each governance activity you improve increases your organization's ability to demonstrate reasonable oversight.

Progress matters far more than perfection.

Start the Conversation

Perhaps the greatest value of this book is its ability to create productive conversations among Boards, executives, cybersecurity leaders, risk professionals, auditors, and legal counsel.

Read chapters together.

Discuss the examples.

Evaluate existing governance practices.

Challenge assumptions.

Ask whether your organization's governance would withstand external scrutiny.

Organizations rarely improve governance because someone purchased a new technology platform.

They improve because leadership begins asking better questions.

Remember the Central Principle

Every chapter in this book points toward a single idea:

Governance is an activity. Evidence is its legacy.

Meetings end.

Policies change.

Leaders retire.

Technologies become obsolete.

But the evidence of governance remains.

When regulators investigate, insurers evaluate claims, auditors perform reviews, or courts examine organizational conduct, that evidence becomes the story your organization tells about its leadership.

This book will help you ensure that it is a story worth telling.

Part I
The Governance Revolution

Chapter 1
The Governance Gap

Most organizations believe they have a cybersecurity problem.

They are wrong.

What they actually have is a governance problem.

That statement may seem counterintuitive. After all, organizations continue to experience ransomware attacks, data breaches, business email compromise, supply chain attacks, insider threats, and a growing list of sophisticated cyber risks. Every day brings headlines describing another organization brought to its knees by a cyber incident.

The natural conclusion is that cybersecurity itself is failing.

Yet that conclusion overlooks a more important reality.

Cybersecurity technologies have never been more capable. Organizations spend billions of dollars annually on firewalls, endpoint protection, identity management, threat intelligence, cloud security, security awareness training, managed detection and response, artificial intelligence, and countless other defensive capabilities. Security teams have access to more tools, more threat data, and more technical expertise than at any point in history.

Despite those investments, the incidents continue.

Why?

Because cybersecurity tools do not govern themselves.

Technology can detect malicious activity. It can block unauthorized access. It can encrypt sensitive information and recover systems from backups. What technology cannot do is establish organizational priorities, determine acceptable levels of risk, allocate resources, resolve competing business objectives, or exercise fiduciary oversight.

Those responsibilities belong to leadership.

That is where the real challenge begins.

Cybersecurity Has Changed

For decades, cybersecurity was viewed primarily as an operational concern.

. . .

Responsibility rested within the information technology department. Success was measured through technical metrics such as patch compliance, firewall effectiveness, antivirus coverage, vulnerability remediation, and system uptime.

Boards rarely discussed cybersecurity unless a major incident occurred.

Executives often viewed cybersecurity as a necessary cost of doing business rather than a strategic capability.

This model made sense when information technology primarily supported business operations.

It no longer reflects reality.

Today, organizations are digital enterprises regardless of their industry.

Hospitals depend on interconnected clinical systems to provide patient care.

Manufacturers rely on automated production systems and global supply chains.

. . .

Financial institutions process millions of digital transactions each day.

Government agencies deliver essential public services electronically.

Nonprofit organizations protect donor information, manage grants, and provide services through increasingly digital platforms.

Technology is no longer a support function.

It is the organization.

Consequently, cybersecurity no longer protects only information systems.

It protects the organization's ability to fulfill its mission.

The Questions Have Changed

Perhaps the most significant shift has occurred after cyber incidents.

Years ago, investigations focused almost exclusively on technical failures.

How did the attacker gain access?

. . .

What vulnerability was exploited?

Which systems were affected?

How much data was compromised?

Those questions remain important.

But they are no longer sufficient.

Today, investigators, regulators, insurers, plaintiffs' attorneys, and shareholders ask a different set of questions.

Did leadership understand the organization's cyber risks?

Was the Board informed?

Were appropriate governance structures established?

Did management report material risks?

Were known deficiencies addressed?

. . .

Who accepted the residual risk?

Can the organization demonstrate that reasonable oversight occurred?

Notice that none of these are technical questions.

They are governance questions.

An organization may deploy world-class cybersecurity technologies and still struggle to answer them.

Conversely, an organization with modest technical capabilities but disciplined governance often demonstrates that leadership exercised reasonable care, made informed decisions, and fulfilled its fiduciary responsibilities.

That distinction matters.

Increasingly, it is the distinction that determines how organizations are judged after an incident.

The Governance Gap

This difference between technical capability and governance maturity is what I call the **governance gap**.

. . .

The Defensible Evidence Framework™

The governance gap exists whenever an organization's cybersecurity program evolves faster than its governance.

Security operations mature.

Governance does not.

Tools multiply.

Oversight remains informal.

Risk assessments are performed.

Risk ownership is unclear.

Security reports are produced.

Boards struggle to interpret them.

Policies exist.

Evidence of oversight does not.

. . .

The organization believes it is becoming more secure because it is acquiring more technology.

In reality, leadership has not improved its ability to govern that technology.

The governance gap widens.

Ironically, organizations often discover this gap only after an incident.

Under normal operating conditions, weak governance may remain invisible for years.

Following a breach, however, every missing policy, undocumented decision, undefined responsibility, and absent Board discussion suddenly becomes significant.

What was once considered an administrative oversight becomes evidence—or the absence of evidence.

Governance Is More Than Compliance

One reason organizations struggle with governance is that they confuse it with compliance.

Compliance asks:

"Did we satisfy the requirement?"

Governance asks:

"Did we make the right decision?"

Compliance measures conformance.

Governance exercises judgment.

Compliance often ends when an audit is complete.

Governance never ends.

Organizations that pursue compliance alone frequently produce impressive collections of policies, procedures, assessments, certifications, and audit reports.

Yet these artifacts do not necessarily demonstrate effective governance.

. . .

A policy that is never discussed by leadership is not evidence of oversight.

A risk assessment that never influences resource allocation is not evidence of governance.

A dashboard reviewed without meaningful discussion is not evidence of informed decision-making.

Governance requires active engagement.

Evidence merely documents that engagement occurred.

The Cost of Poor Governance

Weak governance creates consequences that extend far beyond cybersecurity.

Organizations lose credibility with regulators.

Insurance claims become more difficult to defend.

Audit findings multiply.

Stakeholder confidence erodes.

. . .

Recovery becomes slower because leadership lacks established decision-making structures.

Perhaps most importantly, organizations lose the ability to demonstrate that they acted reasonably.

That word—*reasonably*—deserves careful attention.

The objective of cybersecurity governance is not perfection.

No organization can prevent every attack.

No Board can anticipate every threat.

No executive team can eliminate all risk.

Reasonableness is the appropriate standard.

Reasonable governance means leadership exercised due care.

It understood material risks.

It established appropriate oversight.

. . .

It allocated resources based on informed judgment.

It monitored progress.

It responded to changing conditions.

And it documented those activities as part of normal governance.

That is what investigators, courts, regulators, insurers, and stakeholders increasingly seek to determine.

The Missing Piece

Most governance frameworks describe what organizations should do.

Few explain how organizations prove they did it.

That distinction is the foundation of this book.

Cybersecurity governance should not be judged solely by intentions, policies, or technology investments.

It should be judged by the evidence governance naturally produces.

. . .

When governance becomes part of the organization's operating rhythm, evidence accumulates naturally through Board meetings, committee reports, risk reviews, policy approvals, management briefings, audits, tabletop exercises, corrective actions, and continuous oversight.

Organizations should not create evidence because they anticipate litigation or regulatory scrutiny.

They should create evidence because effective governance leaves a record of informed leadership.

That record becomes invaluable when questions arise months or years later.

Looking Ahead

The chapters that follow introduce a different way of thinking about cybersecurity governance.

Rather than beginning with technology, we begin with leadership.

Rather than focusing on controls, we focus on accountability.

Rather than measuring activity, we evaluate evidence.

This shift may seem subtle.

. . .

It is anything but.

It changes how Boards ask questions.

It changes how executives make decisions.

It changes how security leaders communicate risk.

Most importantly, it changes how organizations demonstrate that they exercised reasonable cybersecurity governance.

Because in today's environment, governance is no longer judged solely by what leaders say they did.

It is judged by what they can prove.

Chapter 2
The Monday Morning Question

Every cyber incident follows two timelines.

The first is technical.

The second is organizational.

The technical timeline begins the moment an attacker gains access to a system. Security teams investigate alerts, contain malicious activity, preserve evidence, eradicate malware, recover systems, and restore operations. Their focus is immediate, tactical, and intensely operational.

The organizational timeline begins later.

Sometimes it begins hours after the incident is discovered.

Sometimes it begins days or weeks later.

. . .

Occasionally it begins months after the systems have been restored.

Its questions are very different.

What happened?

Who knew?

What decisions were made?

Were appropriate people informed?

Did leadership exercise reasonable oversight?

Can the organization demonstrate that oversight?

Those questions are the beginning of what I call **the Monday Morning Question.**

Monday Morning

Imagine arriving at work on a Monday morning.

Your organization experienced a significant cyber incident over the weekend.

. . .

Operations have resumed.

The immediate crisis has stabilized.

Reporters are calling.

Customers are asking questions.

Your cyber insurance carrier has opened a claim.

Outside counsel has been retained.

Regulators have requested information.

The Board has called an emergency meeting.

The organization's technical teams are exhausted but optimistic. They know how the attacker entered the network. They understand the affected systems. Recovery is progressing according to plan.

Then someone asks a deceptively simple question.

"Can we demonstrate that leadership exercised reasonable cybersecurity governance before this incident occurred?"

The room becomes quiet.

No one is asking whether security tools were deployed.

No one is debating firewall configurations.

No one wants to know how many vulnerabilities were patched last quarter.

They want to know whether governance occurred.

That single question changes the conversation.

The Investigation Begins

Contrary to popular belief, cyber investigations rarely remain technical for long.

Every major stakeholder eventually begins evaluating leadership.

. . .

Regulators examine governance practices.

Auditors review oversight processes.

Insurance carriers evaluate whether policy obligations were fulfilled.

Plaintiffs' attorneys look for evidence of negligence.

Investors assess executive judgment.

Boards evaluate management's decisions.

The incident itself becomes only one part of a much larger examination.

The organization is no longer defending its technology.

It is defending its governance.

The Questions That Matter

Although every investigation is different, they tend to converge around a familiar set of questions.

. . .

Did leadership understand the organization's cyber risks?

How were those risks identified?

How frequently were they reviewed?

Were material risks communicated to executive leadership and the Board?

Did the Board receive regular cybersecurity reporting?

Were significant deficiencies discussed?

Were corrective actions assigned?

Who accepted residual risk?

Was that acceptance documented?

Were policies approved and periodically reviewed?

Did management conduct tabletop exercises?

. . .

Were previous audit findings addressed?

Were lessons learned from earlier incidents incorporated into governance?

Notice something important.

These questions are not asking whether an attacker was sophisticated.

They are asking whether leadership was.

Governance Is Easier to Evaluate Than Security

Perfect cybersecurity cannot be measured.

Organizations defend against millions of potential attack paths, thousands of software vulnerabilities, evolving adversaries, and unpredictable human behavior.

There will always be another threat.

Another exploit.

. . .

Another attack.

Governance is different.

Governance leaves evidence.

Policies exist—or they do not.

Board minutes exist—or they do not.

Risk registers are maintained—or they are not.

Committee reports were delivered—or they were not.

Risk acceptance decisions were documented—or they were not.

Tabletop exercises occurred—or they did not.

Leadership discussed cybersecurity—or it did not.

Unlike cybersecurity effectiveness, governance activities can often be objectively demonstrated.

. . .

That is precisely why investigators focus on them.

The Difference Between Knowing and Proving

Many organizations genuinely believe they govern cybersecurity effectively.

Board members remember productive discussions.

Executives recall difficult decisions.

Security leaders remember presenting important risks.

Months later, however, memories become unreliable.

Leadership changes.

Committee memberships rotate.

Key personnel retire.

Documentation becomes incomplete.

Verbal discussions disappear.

. . .

What remains is the evidence.

That distinction is critical.

There is an enormous difference between saying governance occurred and demonstrating that it occurred.

One depends upon recollection.

The other depends upon evidence.

Investigators trust evidence.

The Evidence Problem

Organizations frequently discover they have an evidence problem without realizing they have a governance problem.

The symptoms are familiar.

Cybersecurity reports were presented, but no record exists of Board discussion.

. . .

Risk acceptance occurred through informal conversations rather than documented decisions.

Policies were assumed to be current but had not been reviewed for several years.

Committee charters existed but no longer reflected actual responsibilities.

Management believed accountability was understood, yet no formal assignment of ownership existed.

None of these deficiencies necessarily caused the incident.

They simply make it difficult to demonstrate that leadership governed appropriately before the incident occurred.

The absence of evidence creates uncertainty.

Uncertainty invites scrutiny.

Governance Creates the Organizational Story

Every organization tells a story after a cyber incident.

. . .

Sometimes leadership tells that story through press conferences and public statements.

Sometimes attorneys tell it in court.

Sometimes regulators summarize it in enforcement actions.

Sometimes insurers reconstruct it during claims investigations.

Regardless of who tells the story, it is ultimately supported—or contradicted—by evidence.

Evidence explains what leadership knew.

Evidence demonstrates when decisions were made.

Evidence establishes whether oversight occurred.

Evidence shows whether governance improved over time.

Without that evidence, others will construct the narrative.

Organizations rarely benefit when someone else tells their governance story.

Beyond Compliance

Some readers may wonder whether maintaining this level of documentation simply creates more compliance work.

It does not.

Compliance often asks organizations to retain documentation because regulations require it.

The Defensible Evidence Framework™ takes a different approach.

It asks organizations to govern well enough that meaningful evidence is naturally created during ordinary leadership activities.

Good governance produces evidence.

Evidence is not manufactured after an incident.

It accumulates over months and years through responsible leadership.

That distinction separates governance from documentation.

A Different Way to Think About Cybersecurity

Most cybersecurity books begin with threat actors, malware, vulnerabilities, or security controls.

Those subjects matter.

This book begins somewhere else.

It begins with the realization that every cyber incident eventually becomes a governance investigation.

That realization changes the objective.

Instead of asking only,

> *"How do we prevent attacks?"*

---

leaders begin asking,

> *"If prevention fails, can we demonstrate that we governed responsibly?"*

. . .

The answer to that question has profound implications.

It influences Board agendas.

It shapes executive reporting.

It changes how risk is documented.

It strengthens accountability.

Most importantly, it creates confidence that leadership can demonstrate—not merely claim—that it exercised reasonable cybersecurity governance.

Preparing for the Framework

The Monday Morning Question is not intended to create fear.

It is intended to create clarity.

It provides a practical lens through which every governance activity can be evaluated.

. . .

As you read the remainder of this book, return to this question repeatedly:

If an investigator asked for evidence of this governance activity tomorrow morning, what could we produce?

That simple question becomes the organizing principle for everything that follows.

Because cybersecurity incidents may begin with attackers.

But they almost always end with questions about leadership.

Chapter 3
The Case for Reasonable Cybersecurity Governance

Every organization will experience cybersecurity risk.

Not every organization will experience a catastrophic cyber incident.

No organization can eliminate either.

That reality presents leadership with an uncomfortable truth.

If perfect security is unattainable, what exactly are Boards and executives expected to achieve?

The answer is not perfect cybersecurity.

The answer is **reasonable cybersecurity governance**.

This distinction lies at the heart of modern governance and forms the philosophical foundation of the Defensible Evidence Framework™.

Chasing the Wrong Objective

For years, organizations have been encouraged to pursue stronger security.

More controls.

More monitoring.

More detection.

More compliance.

More technology.

Each of these investments can improve an organization's security posture. None, however, guarantees that a cyber incident will never occur.

In fact, some of the world's most technologically sophisticated organizations have experienced devastating breaches despite investing hundreds of millions of dollars in cybersecurity.

Their failures remind us of an important principle.

. . .

Cybersecurity is not measured by the absence of attacks.

It is measured by an organization's ability to manage risk responsibly.

Governance exists to ensure that management of risk is deliberate, informed, and accountable.

The Meaning of "Reasonable"

The word *reasonable* appears frequently in law, governance, auditing, and risk management because it acknowledges an important limitation.

Leaders cannot control everything.

They cannot predict every threat.

They cannot eliminate every vulnerability.

They cannot foresee every consequence.

Reasonableness recognizes that decisions must be made with incomplete information, finite resources, competing priorities, and evolving risks.

A reasonable Board does not ignore cyber risk.

. . .

A reasonable executive team does not postpone difficult decisions indefinitely.

A reasonable organization does not wait for regulators to dictate its governance practices.

Instead, reasonable leadership asks appropriate questions, evaluates available information, exercises informed judgment, and acts in the best interests of the organization.

Those actions define governance.

Governance Is About Judgment

Technology follows instructions.

Governance exercises judgment.

Security technologies can identify malicious activity, block unauthorized access, encrypt sensitive information, and monitor networks around the clock.

They cannot determine how much risk the organization is willing to accept.

. . .

They cannot balance cybersecurity investments against mission priorities.

They cannot decide whether an operational disruption justifies shutting down a critical business system.

Those decisions belong to people.

More specifically, they belong to leaders entrusted with governing the organization.

Good governance recognizes that cybersecurity decisions are rarely technical decisions alone.

They are business decisions.

Financial decisions.

Legal decisions.

Strategic decisions.

Ethical decisions.

. . .

Technology informs those decisions.

Governance makes them.

The Standard of Care

Every governing body operates according to some standard of care, whether explicitly defined or not.

Boards are expected to act with diligence.

Executives are expected to exercise competent judgment.

Managers are expected to fulfill delegated responsibilities.

Security professionals are expected to provide accurate information and sound recommendations.

Together, these responsibilities create the organization's standard of care.

The question is not whether every decision will prove correct in hindsight.

The question is whether the decision was reasonable based on what leadership knew at the time.

. . .

This distinction is critically important.

Hindsight is perfect.

Governance never is.

Leadership should not be judged by information that became available months after a decision was made.

It should be judged by the quality of the decision-making process that existed when the decision was required.

Decision Quality Matters More Than Decision Outcomes

Organizations sometimes make excellent decisions that produce disappointing outcomes.

Conversely, they occasionally make poor decisions that happen to produce favorable results.

Governance evaluates the quality of the decision-making process, not simply the outcome.

. . .

Consider two organizations that suffer nearly identical ransomware attacks.

The first organization had identified ransomware as a material risk during multiple Board meetings.

Leadership funded improvements based on documented risk assessments.

Recovery capabilities were tested through tabletop exercises.

Residual risks were formally accepted after careful deliberation.

The attack still succeeded.

The second organization also experienced ransomware.

However, cybersecurity had rarely been discussed by leadership.

Risk assessments had not been updated.

Known deficiencies remained unresolved.

Incident response plans had never been exercised.

. . .

The attack also succeeded.

Technically, both organizations experienced similar outcomes.

From a governance perspective, they are fundamentally different.

One can demonstrate reasonable oversight.

The other cannot.

Why Evidence Changes Everything

Without evidence, reasonableness becomes an opinion.

With evidence, it becomes demonstrable.

This is one of the most important concepts in this book.

Organizations often assume that because governance occurred, it can later be reconstructed.

Experience suggests otherwise.

. . .

People remember events differently.

Meeting discussions become blurred.

Context disappears.

Details fade.

Evidence preserves what memory cannot.

It establishes chronology.

It demonstrates deliberation.

It documents accountability.

It records leadership's understanding of risk at specific points in time.

Evidence does not make governance better.

It makes governance demonstrable.

. . .

That distinction becomes enormously important when leadership must explain its decisions months or years after they were made.

Governance Is a Continuous Activity

Reasonable cybersecurity governance is not achieved during annual planning meetings.

It is not created by passing a Board resolution.

It is not established through a single risk assessment.

Governance is continuous.

Every Board meeting contributes to governance.

Every committee discussion contributes to governance.

Every policy approval contributes to governance.

Every risk review contributes to governance.

Every internal audit contributes to governance.

Every tabletop exercise contributes to governance.

. . .

Every corrective action contributes to governance.

Collectively, these activities create an evolving record of leadership.

That record becomes increasingly valuable over time because it demonstrates consistency rather than isolated actions.

Organizations rarely establish reasonableness through a single document.

They establish it through sustained governance.

The Burden of Demonstration

Modern leadership faces an expectation that did not exist to the same degree a decade ago.

It is no longer sufficient to govern responsibly.

Organizations must also demonstrate that they governed responsibly.

This expectation continues to grow.

. . .

Regulators increasingly request documentation rather than assurances.

Cyber insurers evaluate governance maturity before issuing or renewing policies.

Independent auditors examine governance processes alongside technical controls.

Stakeholders expect transparency regarding leadership oversight.

The burden is shifting from assertion to demonstration.

The organizations best prepared for this shift are not necessarily those with the largest cybersecurity budgets.

They are those whose governance consistently produces credible evidence.

A New Definition of Success

Traditional cybersecurity measures success by preventing incidents.

Governance measures success differently.

Success is demonstrated when leadership consistently:

- Understands material cyber risks.
- Establishes clear accountability.
- Makes informed decisions.
- Allocates resources appropriately.
- Validates that governance is functioning.
- Learns from changing conditions.
- Preserves evidence of those activities.

Notice that none of these require perfect security.

All of them require disciplined leadership.

That is an achievable standard.

Looking Ahead

If reasonable governance is the objective, an obvious question follows.

What evidence demonstrates that reasonable governance actually occurred?

Answering that question is the purpose of the Defensible Evidence Framework™.

The framework does not replace existing cybersecurity frameworks.

. . .

It complements them.

It asks a different question.

Not **"What security controls should we implement?"**

But rather,

> **"What evidence should exist if governance is functioning as intended?"**

The answer begins with understanding the framework itself.

That is where we turn next.

Part II
The Defensible Evidence Framework™

Chapter 4
Introducing the Defensible Evidence Framework™

Every governance framework answers a question.

Some answer, *"What risks should we manage?"*

Others answer, *"What controls should we implement?"*

Still others answer, *"How should we organize governance responsibilities?"*

The Defensible Evidence Framework™ answers a different question.

What evidence should exist if an organization has exercised reasonable cybersecurity governance?

That single question fundamentally changes how organizations think about governance.

. . .

Instead of viewing evidence as something assembled after an incident, the framework treats evidence as the natural by-product of good governance. If leadership governs effectively, evidence should accumulate continuously through ordinary governance activities. If evidence is absent, leadership should ask whether governance itself is missing—or merely undocumented.

This distinction is the foundation of the framework.

Why Another Framework?

The cybersecurity profession does not suffer from a shortage of frameworks.

Organizations already have access to outstanding resources, including NIST, ISO/IEC 27001, COBIT, CIS Controls, COSO, CMMC, sector-specific regulations, and numerous industry best practices. Each provides valuable guidance for managing cybersecurity risk.

The Defensible Evidence Framework™ is not intended to replace any of them.

It was never designed to compete with existing standards.

Instead, it complements them.

. . .

Traditional frameworks primarily focus on **what organizations should do.**

The Defensible Evidence Framework™ focuses on **what organizations should be able to demonstrate.**

Those are related—but very different—objectives.

An organization may implement every recommended security control yet struggle to demonstrate that leadership exercised appropriate oversight.

Conversely, an organization may possess excellent governance evidence while still identifying opportunities to improve its technical controls.

The two perspectives strengthen one another.

One improves security.

The other improves defensibility.

From Activities to Evidence

Most organizations measure cybersecurity through activities.

. . .

Policies approved.

Vulnerabilities remediated.

Systems patched.

Employees trained.

Incidents resolved.

Audits completed.

These activities are important, but activity alone does not demonstrate governance.

Governance requires leadership.

Leadership produces decisions.

Decisions produce evidence.

That relationship is central to this book.

. . .

Consider a quarterly Board cybersecurity briefing.

The briefing itself is an activity.

The Board's discussion, questions, requested follow-up actions, documented decisions, and subsequent monitoring become evidence of governance.

Similarly, a tabletop exercise is an activity.

The exercise objectives, participant roster, observations, corrective actions, executive review, and follow-up validation collectively become evidence.

The framework encourages organizations to stop thinking solely about activities and begin thinking about the evidence those activities naturally produce.

Governance Should Leave a Trail

Imagine attempting to reconstruct an organization's cybersecurity governance five years after a significant breach.

Leadership has changed.

Board members have rotated.

. . .

Committee structures have evolved.

Key employees have retired.

The only reliable source of truth is the governance record.

That record should answer important questions.

How was cybersecurity governed?

What risks were considered?

What priorities were established?

How did leadership respond to emerging threats?

How did governance improve over time?

Organizations that govern well leave a trail.

Organizations that govern poorly often leave only assumptions.

. . .

The Defensible Evidence Framework™ is designed to ensure that the governance trail exists before anyone asks for it.

The Five Evidence Domains

The framework organizes governance evidence into five complementary domains.

Together, they describe the complete lifecycle of governance.

Governance Evidence

Governance begins with organizational direction.

This domain captures the evidence that leadership established authority, defined responsibilities, approved policies, delegated oversight, and created the structures necessary for governance to function.

Governance Evidence answers questions such as:

- Who governs cybersecurity?
- What authority has been established?
- How is accountability assigned?
- What expectations has leadership communicated?

Without Governance Evidence, organizations struggle to demonstrate that cybersecurity oversight was ever formally established.

Risk Evidence

Governance exists to manage risk.

Risk Evidence documents how leadership identifies, evaluates, prioritizes, accepts, mitigates, transfers, and monitors cybersecurity risk.

This domain demonstrates that cyber risk is governed as enterprise risk rather than merely managed as a technical concern.

Risk Evidence answers questions including:

- What risks were known?
- How significant were they?
- Who evaluated them?
- What decisions were made?
- Which risks were accepted?

Risk Evidence transforms cybersecurity from an operational discussion into a leadership decision process.

Oversight Evidence

Governance requires continuous oversight.

Oversight Evidence demonstrates that leadership remained engaged over time through Board reporting, committee meetings, executive briefings, performance reviews, corrective action tracking, and governance monitoring.

This domain answers questions such as:

- Did leadership receive meaningful information?
- Were concerns discussed?
- Were management actions monitored?
- Did oversight continue over time?

Oversight Evidence demonstrates that governance was active rather than passive.

Operational Evidence

Governance ultimately influences operations.

Operational Evidence captures the implementation of governance decisions through incident response exercises, awareness programs, operational reporting, vendor oversight, continuity planning, resilience testing, and other management activities.

Unlike Governance Evidence, which focuses on leadership direction, Operational Evidence demonstrates that governance decisions were translated into organizational action.

It answers questions such as:

- Were governance decisions implemented?
- Were plans exercised?
- Were personnel prepared?
- Did operational capabilities improve?

Operational Evidence connects governance to execution.

Assurance Evidence

Good governance verifies.

It does not assume.

Assurance Evidence documents independent validation through internal audits, external assessments, penetration testing, maturity reviews, compliance evaluations, corrective action verification, and similar activities.

This domain provides leadership with confidence that governance is functioning as intended.

It answers questions including:

- Were governance activities independently validated?
- Were deficiencies identified?
- Were corrective actions completed?
- Did leadership receive objective assurance?

. . .

Without Assurance Evidence, organizations rely heavily on optimism.

With Assurance Evidence, they rely on verification.

Why These Five Domains Matter

Although each evidence domain serves a distinct purpose, none stands alone.

Governance establishes direction.

Risk informs decisions.

Oversight monitors progress.

Operations implement decisions.

Assurance validates effectiveness.

Together, they create a continuous cycle of governance.

Weakness in any domain eventually affects the others.

. . .

An organization with excellent Operational Evidence but weak Oversight Evidence may demonstrate technical competence without proving executive engagement.

Strong Governance Evidence without Assurance Evidence may show that leadership established expectations but never confirmed they were achieved.

The strength of the framework lies in its balance.

Each domain supports the others.

Together, they produce a comprehensive picture of reasonable governance.

A Framework That Fits Existing Governance

One of the framework's greatest strengths is its flexibility.

It does not require organizations to redesign their governance structures.

It does not mandate new committees.

It does not prescribe specific reporting formats.

. . .

It does not require abandoning existing frameworks or certifications.

Instead, it overlays existing governance activities and asks a simple question:

What evidence should naturally exist if this governance activity occurred?

Organizations frequently discover that much of the necessary evidence already exists.

The challenge is not always creating new governance.

Often, it is organizing, preserving, and strengthening the evidence generated by governance already taking place.

Beyond Cybersecurity

Although this book focuses on cybersecurity, the underlying principles extend far beyond cyber risk.

Every governance discipline produces evidence.

Financial governance.

. . .

Privacy governance.

Artificial intelligence governance.

Operational governance.

Environmental governance.

Corporate governance.

The framework is therefore not limited to cybersecurity.

Cybersecurity simply provides one of the clearest and most urgent applications because cyber incidents routinely trigger intense external scrutiny.

The principles described throughout this book are ultimately principles of leadership accountability.

The Central Idea

The Defensible Evidence Framework™ can be summarized in a single statement:

Effective governance naturally creates defensible evidence.

. . .

Organizations should not create evidence because they fear litigation.

They should not create evidence because regulators may someday ask for it.

They should create evidence because disciplined governance leaves a reliable record of informed leadership.

When governance becomes part of an organization's operating rhythm, evidence ceases to be an administrative burden.

It becomes the legacy of responsible leadership.

The remainder of this book explores each evidence domain in depth, demonstrates the types of evidence that matter most, and provides practical guidance for building governance that is not only effective today, but demonstrably reasonable tomorrow.

THE DEFENSIBLE EVIDENCE FRAMEWORK™ AT A GLANCE

A structured approach to demonstrate reasonable cybersecurity governance through evidence.

GOVERNANCE

1. GOVERNANCE EVIDENCE

Establishes the foundation for cybersecurity governance by defining structures, roles, responsibilities, and policies.

KEY EVIDENCE

- Governance policies and charters
- Roles and responsibilities
- Board and committee mandates
- Governance calendar

RISK

2. RISK EVIDENCE

Demonstrates that cyber risks are identified, assessed, prioritized, and managed within the organization's risk tolerance.

KEY EVIDENCE

- Risk assessments and registers
- Risk appetite and tolerance
- Risk acceptance decisions
- Third-party risk evaluations

OVERSIGHT

3. OVERSIGHT EVIDENCE

Shows that leadership and the Board receive meaningful information and actively exercise informed oversight.

KEY EVIDENCE

- Board and committee minutes
- Executive reports and dashboards
- Decisions and action items
- Follow-up and accountability

OPERATIONS

4. OPERATIONAL EVIDENCE

Demonstrates that cybersecurity plans, processes, and capabilities are established, tested, and executed effectively.

KEY EVIDENCE

- Incident response plans and reports
- Tabletop exercises and lessons learned
- Business continuity and DR testing
- Corrective action management

ASSURANCE

5. ASSURANCE EVIDENCE

Provides independent validation that governance and risk management activities are effective and continuously improved.

KEY EVIDENCE

- Audit and assessment reports
- Management responses
- Remediation tracking
- Continuous improvement metrics

DEFENSIBLE GOVERNANCE

When the Monday Morning Question is asked, your organization has the evidence to answer.

DOCUMENTED | ACCOUNTABLE | TRANSPARENT | DEFENSIBLE

CLARITY. ACCOUNTABILITY. DEFENSIBILITY.

Chapter 5
Governance Evidence

Every organization generates evidence.

Invoices become evidence of financial transactions.

Contracts become evidence of legal agreements.

Personnel records become evidence of employment actions.

Likewise, governance produces evidence.

Or at least it should.

One of the most common misconceptions about governance is that it exists primarily in discussions, experience, or institutional knowledge. Organizations often believe they have exercised effective oversight because leaders remember important conversations, difficult decisions, or strategic debates.

. . .

The problem is that governance cannot rely on memory.

It must rely on evidence.

Governance Evidence forms the first and most foundational domain of the Defensible Evidence Framework™ because it answers the most fundamental question an investigator, regulator, insurer, or auditor can ask:

How was cybersecurity governance established?

Without Governance Evidence, every other evidence domain rests on an uncertain foundation.

Governance Begins Before the First Risk

Organizations often think governance begins when they perform a risk assessment.

It begins much earlier.

Governance begins the moment leadership establishes authority.

Before risks are identified...

. . .

Before controls are implemented...

Before incidents occur...

Someone must decide:

Who governs cybersecurity?

Who owns cyber risk?

Who has decision-making authority?

How is oversight performed?

How are responsibilities delegated?

Those decisions establish the governance system.

Governance Evidence documents that those decisions occurred.

Governance Is Organizational Architecture

An architect begins with structural drawings before construction starts.

. . .

Similarly, governance begins with organizational architecture.

Leadership establishes the framework within which cybersecurity decisions will be made.

This architecture includes:

- Board responsibilities
- Committee responsibilities
- Executive accountability
- Reporting relationships
- Decision authorities
- Escalation paths
- Policy ownership
- Governance calendars
- Oversight processes

Without organizational architecture, cybersecurity becomes reactive.

People make decisions based upon urgency rather than authority.

Responsibilities become blurred.

Critical issues fall between organizational boundaries.

. . .

Governance Evidence demonstrates that leadership intentionally designed how cybersecurity would be governed.

Governance Evidence Answers Foundational Questions

Governance Evidence exists to answer questions that arise long before an incident occurs.

For example:

Who is responsible for cybersecurity governance?

Does the Board formally oversee cybersecurity?

Which committee receives cyber reports?

Who approves cybersecurity policy?

Who accepts residual cyber risk?

Who determines reporting frequency?

How are material risks escalated?

. . .

Who independently validates governance?

These questions rarely require technical expertise.

They require organizational clarity.

The answers should never depend upon verbal explanations.

They should already exist within the organization's governance evidence.

What Constitutes Governance Evidence?

Governance Evidence consists primarily of documents that establish authority, responsibility, and organizational intent.

Examples include:

- Board-approved cybersecurity governance policies
- Committee charters
- Board resolutions
- Governance frameworks
- Organizational charts
- Executive accountability assignments
- Delegations of authority
- Governance calendars
- Policy approval records
- Policy review histories
- Governance committee meeting schedules

- Board education plans
- Governance operating procedures

Collectively, these artifacts tell the story of how leadership intended cybersecurity governance to function.

Notice that these documents are strategic rather than operational.

They describe leadership.

Not implementation.

Governance Intent Matters

Policies are often criticized because they do not directly improve cybersecurity.

That criticism misunderstands their purpose.

Policies do not secure systems.

People do.

Policies establish expectations.

. . .

Governance establishes accountability.

Consider a Board-approved cybersecurity governance policy.

The policy itself will never stop ransomware.

However, it establishes who is responsible for governing cyber risk, defines reporting expectations, assigns oversight responsibilities, and creates organizational accountability.

Those governance decisions influence everything that follows.

The policy becomes evidence that leadership intentionally established expectations rather than relying on assumptions.

Intent matters.

Organizations that clearly communicate expectations generally govern more effectively than organizations that assume everyone already understands them.

The Importance of Formal Authority

Many organizations operate successfully for years through informal leadership relationships.

. . .

People know whom to call.

Decisions happen quickly.

Responsibilities seem obvious.

Until leadership changes.

When experienced executives retire, Boards rotate, organizations merge, or crises occur, informal governance frequently begins to fail.

People discover that everyone understood the process differently.

Formal governance prevents that uncertainty.

Formal authority establishes consistency independent of personalities.

It survives leadership transitions.

It preserves organizational continuity.

Governance Evidence demonstrates that authority belongs to the organization—not merely to individuals.

Policies Are Governance Instruments

Organizations frequently view cybersecurity policies as compliance documents.

In reality, policies are governance instruments.

A governance policy communicates leadership's decisions regarding acceptable organizational behavior.

It answers questions such as:

What is required?

Who is accountable?

Who has authority?

What exceptions are permitted?

How will compliance be monitored?

What consequences exist for noncompliance?

. . .

These questions extend well beyond cybersecurity.

They define organizational expectations.

Every approved policy therefore becomes evidence that leadership exercised governance.

Committee Charters Matter More Than Most Organizations Realize

Many organizations maintain committee charters because corporate governance practices require them.

Few appreciate their evidentiary value.

A committee charter establishes:

- Purpose
- Scope
- Authority
- Membership
- Reporting responsibilities
- Decision rights
- Meeting frequency

During a governance review, the charter demonstrates that leadership intentionally established oversight rather than relying upon informal discussions.

. . .

When supported by meeting minutes and regular reporting, committee charters become powerful evidence of sustained governance.

Governance Calendars Create Predictability

Effective governance follows a rhythm.

Boards meet regularly.

Committees meet regularly.

Risk reports are delivered regularly.

Policies are reviewed regularly.

Training occurs regularly.

Governance calendars formalize this cadence.

Rather than depending upon ad hoc discussions, governance becomes predictable.

Predictability produces consistency.

. . .

Consistency produces evidence.

Organizations that govern consistently rarely struggle to demonstrate oversight because governance activities occur according to an established schedule.

Governance Evidence Ages

One of the most overlooked aspects of Governance Evidence is that it has a useful life.

Policies written seven years ago but never reviewed provide limited assurance.

Committee charters that no longer reflect organizational responsibilities become misleading.

Organizational charts that omit current leadership undermine credibility.

Evidence should demonstrate active governance—not historical governance.

This is why review histories are often as valuable as the documents themselves.

. . .

A policy approved in 2022 tells part of the story.

A policy reviewed annually since 2022 tells a much stronger one.

Governance is continuous.

Its evidence should reflect that continuity.

Common Governance Evidence Gaps

Organizations conducting their first Governance Evidence review often discover recurring weaknesses.

Common gaps include:

Policies without documented approval.

Policies lacking periodic review.

Committee charters that no longer reflect actual practice.

Undefined executive accountability.

Ambiguous risk ownership.

. . .

Missing Board education records.

Governance responsibilities scattered across multiple documents.

Informal governance processes that exist only through custom and practice.

None of these deficiencies necessarily indicate poor leadership.

They often reflect governance that evolved organically over many years.

The purpose of Governance Evidence is not to criticize these organizations.

It is to strengthen them.

Building Better Governance Evidence

Organizations do not strengthen Governance Evidence by creating more documents.

They strengthen it by improving governance.

When leadership clearly defines authority...

. . .

Documents naturally follow.

When Boards formally approve governance policies...

Evidence naturally follows.

When committees meet according to established charters...

Evidence naturally follows.

When governance becomes disciplined...

Evidence accumulates naturally.

This distinction cannot be overstated.

The objective is never paperwork.

The objective is better governance.

Paperwork merely documents it.

Governance Evidence as the Foundation

Every remaining evidence domain depends upon Governance Evidence.

Risk Evidence assumes someone has authority to evaluate and accept risk.

Oversight Evidence assumes governance structures exist.

Operational Evidence assumes leadership has established expectations.

Assurance Evidence assumes someone has defined what success looks like.

Governance Evidence therefore becomes the cornerstone of the Defensible Evidence Framework™.

It answers the first question every external reviewer eventually asks:

How did leadership establish cybersecurity governance?

If that question cannot be answered confidently, the remaining evidence domains become increasingly difficult to defend.

Because before leadership can demonstrate that it governed cybersecurity well, it must first demonstrate that governance itself was intentionally established.

Chapter 6
Risk Evidence

Governance without risk is administration.

Risk without governance is gambling.

The purpose of cybersecurity governance is not to eliminate risk. It is to understand risk well enough that leadership can make informed, deliberate, and defensible decisions about it.

That is the role of Risk Evidence.

Risk Evidence documents how an organization identifies, evaluates, prioritizes, communicates, and responds to cybersecurity risk. More importantly, it demonstrates that leadership—not simply technical staff—participated in those decisions.

When investigators examine governance following a cyber incident, they are rarely interested in every vulnerability that ever existed.

They want to understand something much more important.

Did leadership know about the risks that mattered?

Risk Evidence answers that question.

Risk Exists Before an Incident

One of the greatest misconceptions in cybersecurity is that risk becomes important only after something goes wrong.

In reality, every significant cyber incident begins as an unmanaged—or insufficiently managed—risk.

Long before ransomware encrypts servers...

Long before customer information is stolen...

Long before operations are disrupted...

Someone understood at least part of the organization's exposure.

Perhaps it appeared in a vulnerability assessment.

. . .

Perhaps it was identified during a tabletop exercise.

Perhaps an internal audit noted the issue.

Perhaps management discussed it during a quarterly review.

Perhaps a third-party assessment highlighted the concern.

The question is not whether the risk existed.

The question is what leadership did with that knowledge.

Risk Is Information for Decision-Making

Many organizations treat risk registers as inventories.

Lists of vulnerabilities.

Lists of concerns.

Lists of projects awaiting funding.

That approach misses their true purpose.

. . .

Risk exists to support decision-making.

Every identified risk should answer four fundamental questions.

What could happen?

How likely is it?

What would the consequences be?

What should leadership do about it?

If a risk assessment cannot support those decisions, it has become an administrative exercise rather than a governance instrument.

Risk Evidence documents the quality of that decision-making process.

The Difference Between Technical Risk and Enterprise Risk

Technical teams naturally evaluate cyber risk through technical lenses.

Unpatched software.

. . .

Weak authentication.

Cloud misconfigurations.

Legacy operating systems.

These issues matter.

Boards, however, govern organizations—not technologies.

Enterprise Risk asks different questions.

Could this disrupt our mission?

Could this interrupt critical services?

Could this violate legal obligations?

Could this expose sensitive information?

Could this damage stakeholder trust?

Could this materially affect financial performance?

. . .

The same technical vulnerability may represent vastly different levels of enterprise risk depending upon where it exists.

A vulnerable laboratory workstation differs dramatically from a vulnerable patient care system.

An unpatched development server carries different implications than an unpatched payment processing platform.

Risk Evidence demonstrates that leadership evaluated business consequences—not merely technical severity.

Risk Registers Tell a Story

Every mature governance program maintains some form of risk register.

Unfortunately, many organizations treat it as a static spreadsheet updated only before audits.

A governance-oriented risk register tells an evolving story.

It shows when risks were identified.

Who evaluated them.

. . .

How severity changed over time.

What mitigation activities occurred.

Which risks were accepted.

Who approved those decisions.

When risks were reviewed again.

Viewed this way, the risk register becomes more than a management tool.

It becomes a historical record of governance.

Months or years later, leadership can reconstruct exactly how significant risks were understood and managed.

That chronology is invaluable.

Risk Appetite Creates Context

Not every cybersecurity risk deserves the same response.

. . .

Some require immediate mitigation.

Others can be monitored.

Still others may be accepted because the cost of eliminating them exceeds the benefit.

These decisions should never occur in isolation.

They should reflect the organization's risk appetite.

Risk appetite defines the level of uncertainty leadership is willing to accept while pursuing organizational objectives.

Without an established risk appetite, risk decisions become inconsistent.

Different executives reach different conclusions.

Departments apply different standards.

Priorities shift depending upon personalities rather than governance.

. . .

Risk Evidence should demonstrate that risk decisions align with organizational risk appetite rather than individual preferences.

Risk Acceptance Is a Governance Decision

Perhaps no governance activity is more misunderstood than risk acceptance.

Some organizations treat accepting risk as doing nothing.

Others assume technical staff possess authority to accept significant organizational risk.

Both assumptions are dangerous.

Risk acceptance is an executive decision.

It acknowledges that leadership understands the residual exposure after reasonable mitigation and intentionally chooses to proceed.

That decision should never be accidental.

Nor should it be invisible.

Strong Risk Evidence includes documentation showing:

- The risk being accepted.
- The potential business impact.
- Alternatives considered.
- Residual exposure.
- The individual or governing body authorized to accept the risk.
- The date of acceptance.
- Conditions requiring future review.

This documentation does not assign blame.

It demonstrates informed judgment.

Risk Decisions Should Age Gracefully

Cyber risk changes continuously.

New threats emerge.

Business priorities evolve.

Technologies are replaced.

Regulations expand.

. . .

Consequently, yesterday's acceptable risk may become today's unacceptable exposure.

Risk Evidence should therefore demonstrate periodic reassessment.

Leadership should periodically ask:

Has this risk changed?

Have new controls reduced exposure?

Have business impacts increased?

Does our original acceptance remain appropriate?

Evidence showing thoughtful reevaluation often proves as valuable as the original decision.

It demonstrates that governance remained active rather than static.

Third-Party Risk Is Governance Risk

Few organizations operate independently.

Cloud providers.

. . .

Managed service providers.

Software vendors.

Payroll processors.

Healthcare partners.

Financial institutions.

Critical suppliers.

Every external dependency introduces cyber risk beyond the organization's direct control.

Leadership cannot outsource accountability simply because it outsourced a service.

Risk Evidence should therefore include documentation demonstrating that significant third-party relationships were evaluated before contracts were signed and periodically reassessed thereafter.

. . .

Vendor questionnaires.

Independent assessments.

Contract reviews.

Service-level agreements.

Business continuity evaluations.

Risk acceptance decisions.

These artifacts collectively demonstrate that leadership governed third-party exposure rather than assuming vendor responsibility equated to organizational protection.

Communicating Risk to Leadership

One of the greatest failures in cybersecurity governance occurs when technical risks are never translated into business language.

A Board should never need to interpret Common Vulnerability Scoring System values or analyze exploit chains.

It should understand business consequences.

. . .

Effective Risk Evidence therefore includes executive-level reporting that explains:

What the risk means.

Why it matters.

Potential organizational impact.

Available response options.

Recommended action.

Decision required.

Risk communication is not simplification.

It is translation.

Without translation, governance becomes impossible.

Common Risk Evidence Gaps

Organizations reviewing their Risk Evidence often discover familiar weaknesses.

. . .

Risk registers maintained only for compliance.

Undefined risk ownership.

Undocumented risk acceptance.

Technical risks lacking business impact analysis.

No linkage between risk assessments and investment decisions.

Risk reports that describe problems but request no decisions.

Inconsistent criteria for evaluating severity.

Third-party risks evaluated only during procurement.

These deficiencies rarely indicate poor cybersecurity.

They indicate incomplete governance.

The objective is not merely to identify risks.

. . .

It is to govern them.

Building Better Risk Evidence

Improving Risk Evidence does not require creating more reports.

It requires improving the quality of leadership decisions.

When risks are evaluated consistently...

Evidence improves.

When business consequences accompany technical findings...

Evidence improves.

When ownership is clearly assigned...

Evidence improves.

When acceptance decisions are documented...

Evidence improves.

. . .

When risks are reviewed over time...

Evidence improves.

The evidence reflects the governance.

Never the other way around.

Risk Evidence as Proof of Leadership

Every cyber incident invites a fundamental question.

> **Did leadership understand the organization's exposure before the incident occurred?**

Risk Evidence provides the answer.

Not through assumptions.

Not through recollections.

Not through post-incident explanations.

. . .

Through contemporaneous records demonstrating that leadership identified risks, evaluated their significance, considered available options, made informed decisions, and revisited those decisions as circumstances evolved.

That is what reasonable governance looks like.

It does not promise perfect outcomes.

It demonstrates disciplined leadership.

And in an environment where cyber risk can never be eliminated, disciplined leadership is one of the strongest forms of evidence an organization can produce.

Chapter 7
Oversight Evidence

Governance without oversight is delegation.

Delegation without oversight is abdication.

This distinction lies at the heart of effective cybersecurity governance.

Boards appropriately delegate the execution of cybersecurity to executive leadership. Executive leadership delegates operational responsibilities to management. Management relies upon security professionals, technology teams, and business units to implement the organization's cybersecurity program.

None of these delegations transfers accountability.

The Board remains accountable for oversight.

Executive leadership remains accountable for execution.

. . .

Management remains accountable for results.

Oversight Evidence demonstrates that these responsibilities were actively exercised rather than simply assigned.

It answers one of the most important questions an external reviewer can ask:

How did leadership know governance was working?

Oversight Is Not Observation

Many organizations confuse oversight with awareness.

Receiving a quarterly cybersecurity presentation is awareness.

Reading a dashboard is awareness.

Listening to an executive briefing is awareness.

Oversight requires something more.

It requires engagement.

. . .

Boards exercise oversight when they ask questions.

Challenge assumptions.

Request clarification.

Direct management to investigate concerns.

Require corrective action.

Follow up on previous commitments.

Monitor progress over time.

Without these activities, reporting becomes informational rather than governable.

Oversight Evidence documents that leadership did more than receive information.

It demonstrates that leadership acted upon it.

The Rhythm of Governance

Effective oversight follows a predictable rhythm.

Cybersecurity should never appear on the Board agenda only after a breach.

Instead, oversight becomes part of the organization's normal governance cadence.

Quarterly Board reports.

Committee briefings.

Executive risk reviews.

Policy reviews.

Audit updates.

Tabletop exercise reports.

Corrective action reviews.

. . .

Annual governance assessments.

Each activity contributes another piece of the governance record.

Viewed individually, these events may appear routine.

Viewed collectively, they demonstrate sustained leadership engagement.

Consistency is one of the strongest indicators of governance maturity.

Organizations that govern consistently rarely struggle to demonstrate oversight.

Their evidence accumulates naturally.

Board Minutes Tell a Story

Few governance artifacts receive more scrutiny following a cyber incident than Board and committee minutes.

Unfortunately, many organizations misunderstand their purpose.

Board minutes are not intended to become transcripts.

. . .

Nor should they attempt to capture every conversation.

Their purpose is to document governance.

Effective minutes demonstrate:

The matters considered.

The information presented.

The questions raised.

The decisions made.

The actions assigned.

The follow-up expected.

The objective is not to prove that every Board member understood every technical detail.

The objective is to demonstrate that leadership exercised informed oversight.

. . .

Good Board minutes tell the story of governance.

Poor Board minutes merely record attendance.

Dashboards Do Not Govern

Cybersecurity dashboards have become increasingly sophisticated.

Executives receive charts.

Trend lines.

Heat maps.

Performance metrics.

Risk indicators.

Control scores.

These tools provide valuable information.

But dashboards do not govern organizations.

. . .

Leaders do.

Oversight Evidence therefore extends beyond the dashboard itself.

It includes the discussion surrounding the dashboard.

Which trends concerned leadership?

Which metrics prompted questions?

What actions were requested?

Which issues required additional reporting?

How were unresolved concerns tracked?

A dashboard becomes evidence only when leadership uses it to exercise governance.

Otherwise, it is simply data.

Asking Better Questions

One of the Board's greatest responsibilities is asking questions that management may not ask itself.

. . .

Not because management is negligent.

Because governance requires an independent perspective.

Effective oversight questions include:

Which cyber risks have changed since our last meeting?

What assumptions concern you most?

What keeps you awake at night?

Which recommendations have not yet been funded?

Where do you believe our governance is weakest?

What would concern an external investigator?

What evidence would demonstrate our oversight if an incident occurred tomorrow?

. . .

Questions like these move governance beyond reporting.

They encourage critical thinking.

They reveal assumptions.

They expose uncertainty.

Most importantly, they create evidence that leadership exercised diligence.

Oversight Requires Follow-Through

A surprising number of governance failures occur not because leadership ignored problems, but because it failed to verify that corrective actions were completed.

Consider a common scenario.

Internal Audit identifies weaknesses in privileged access management.

Management agrees to corrective actions.

The Board receives the report.

. . .

Months later, another briefing occurs.

The issue receives no further discussion.

Did the organization resolve the problem?

Perhaps.

Perhaps not.

Without documented follow-up, oversight ends where governance should continue.

Oversight Evidence should demonstrate the complete lifecycle of governance.

Issue identified.

Leadership informed.

Corrective action assigned.

Progress monitored.

. . .

Resolution verified.

Only then is the governance cycle complete.

Escalation Is Evidence of Healthy Governance

Organizations often fear escalation because it appears to signal failure.

Healthy governance views escalation differently.

Escalation demonstrates that governance is functioning.

Employees raise concerns.

Management communicates material issues.

Executives notify the Board.

The Board exercises oversight.

Appropriate decisions follow.

None of these actions indicate organizational weakness.

. . .

They indicate organizational transparency.

Oversight Evidence should therefore include documentation showing that escalation mechanisms were used appropriately.

An organization where nothing is ever escalated may not have fewer problems.

It may simply have fewer conversations.

Oversight Must Include Bad News

One of the greatest threats to governance is optimism bias.

Executives naturally prefer reporting success.

Projects completed.

Controls implemented.

Threats blocked.

Compliance achieved.

. . .

These accomplishments deserve recognition.

They are not sufficient.

Boards govern most effectively when they understand what is not working.

Failed penetration tests.

Unresolved audit findings.

Delayed remediation.

Staffing shortages.

Technology debt.

Vendor deficiencies.

Emerging threats.

Material uncertainties.

. . .

Organizations that consistently report only positive news often create the illusion of effective governance.

Oversight Evidence should demonstrate that leadership received balanced information.

Good governance tolerates uncomfortable conversations.

Poor governance avoids them.

Oversight Is Continuous

Governance is not a quarterly event.

It is an ongoing leadership discipline.

Between formal meetings, oversight continues through:

Executive communications.

Committee updates.

Corrective action monitoring.

. . .

Risk reviews.

Incident briefings.

Emerging threat discussions.

Policy revisions.

Audit follow-up.

Collectively, these activities demonstrate that governance remained active between Board meetings rather than beginning and ending with scheduled presentations.

Common Oversight Evidence Gaps

Organizations evaluating Oversight Evidence frequently discover similar weaknesses.

Cybersecurity appears on Board agendas only after incidents.

Reporting focuses exclusively on technical activity.

Minutes document presentations but not discussion.

. . .

Corrective actions lack documented follow-up.

Dashboards measure operations rather than governance.

Committees receive information without requesting decisions.

Board education occurs infrequently.

Escalation thresholds remain undefined.

These deficiencies rarely result from poor intentions.

More often, they reflect governance practices that evolved informally over time.

Recognizing these gaps is the first step toward strengthening oversight.

Building Better Oversight Evidence

Improving Oversight Evidence begins with improving governance conversations.

Boards should expect reporting that informs decisions rather than merely describing activity.

. . .

Management should present alternatives rather than conclusions.

Committees should document not only what they heard but how they responded.

Corrective actions should remain visible until independently verified.

Questions should be encouraged.

Follow-up should be expected.

Governance should be continuous.

When these practices become routine, evidence accumulates naturally.

No additional paperwork is required.

Only better governance.

Oversight Demonstrates Fiduciary Leadership

Ultimately, Oversight Evidence demonstrates something larger than cybersecurity.

. . .

It demonstrates stewardship.

Stakeholders expect Boards to govern responsibly.

Employees expect leadership to protect organizational resilience.

Customers expect responsible oversight of their information.

Regulators expect informed decision-making.

Investors expect disciplined governance.

Oversight Evidence tells the story of how leadership fulfilled those expectations.

Not through technical expertise.

Not through operational execution.

But through thoughtful, informed, and continuous oversight.

. . .

That is the essence of fiduciary leadership.

And it is one of the strongest indicators that cybersecurity governance is functioning as intended.

Chapter 8
Operational Evidence

Governance exists to influence action.

A Board may approve policies.

Executives may establish priorities.

Committees may review risks.

Auditors may issue recommendations.

None of these activities improve cybersecurity on their own.

Organizations become more resilient only when governance decisions are translated into operational reality.

Operational Evidence demonstrates that translation.

. . .

It provides tangible proof that leadership's decisions moved beyond discussion and became part of the organization's daily operations.

This distinction is critical.

Governance establishes direction.

Operations execute that direction.

Operational Evidence connects the two.

Governance Must Produce Results

An organization may possess exemplary governance documentation.

Board minutes are complete.

Policies are current.

Committee charters are well defined.

Risk registers are meticulously maintained.

Yet if those governance activities never influence operational behavior, governance has failed.

. . .

Governance is not measured by the quality of its documentation.

It is measured by the quality of its outcomes.

Operational Evidence demonstrates that governance produced action.

It answers questions such as:

- Were approved policies implemented?
- Were identified risks mitigated?
- Were assigned responsibilities fulfilled?
- Were personnel prepared?
- Were governance decisions reflected in operational practice?

Without Operational Evidence, leadership cannot demonstrate that governance actually changed the organization.

Operations Reveal Organizational Culture

Operational Evidence often reveals more about an organization than its governance documents.

Policies describe expectations.

. . .

Operations demonstrate commitment.

Organizations frequently publish comprehensive cybersecurity policies while failing to implement basic security practices.

Conversely, mature organizations consistently align operational behavior with governance expectations.

Employees understand reporting procedures.

Managers conduct periodic reviews.

Incident response teams follow established playbooks.

Vendors are monitored.

Security awareness becomes part of normal business operations.

Governance is no longer theoretical.

It becomes organizational culture.

Operational Evidence documents that transformation.

Incident Response Demonstrates Preparedness

Few operational activities receive greater scrutiny following a cyber incident than incident response.

Not because organizations are expected to prevent every attack.

Because they are expected to prepare for them.

Incident response plans represent governance intent.

Their execution produces Operational Evidence.

Strong Operational Evidence includes:

- Incident response plans.
- Defined roles and responsibilities.
- Escalation procedures.
- Communication plans.
- Decision logs.
- Situation reports.
- After-action reviews.
- Lessons learned.
- Corrective action plans.

Collectively, these artifacts demonstrate that leadership anticipated the possibility of a cyber incident and prepared the organization to respond effectively.

Preparation is evidence of responsible governance.

Tabletop Exercises Create Confidence

Organizations often view tabletop exercises as training events.

They are much more than that.

A well-designed tabletop exercise validates governance.

It demonstrates whether:

Decision authority is understood.

Escalation paths function.

Communication plans are realistic.

Executive leadership understands its responsibilities.

. . .

The Board receives appropriate information.

Policies support decision-making.

Recovery expectations are achievable.

Most importantly, tabletop exercises generate evidence before a real crisis occurs.

Exercise objectives.

Participant rosters.

Facilitator observations.

Executive discussions.

Identified weaknesses.

Corrective actions.

Follow-up validation.

. . .

These records become some of the strongest Operational Evidence an organization can produce.

They demonstrate that leadership practiced governance rather than assuming it would function during a crisis.

Security Awareness Is Governance in Action

Cybersecurity awareness is frequently viewed as an information technology initiative.

It is actually a governance activity.

Every employee influences organizational risk.

Every employee handles information.

Every employee makes decisions affecting cybersecurity.

Governance therefore extends beyond executives and technical teams.

Operational Evidence should demonstrate that the organization actively prepares its workforce.

. . .

Evidence may include:

Training completion records.

Role-based education.

Executive briefings.

Board education.

Phishing simulation results.

Policy acknowledgments.

Security awareness campaigns.

Improvement metrics.

The objective is not merely to satisfy compliance requirements.

It is to demonstrate that leadership intentionally fostered a culture supporting responsible cybersecurity behavior.

Vendor Oversight Extends Governance

Modern organizations depend upon dozens—or hundreds—of third parties.

Cloud providers.

Software vendors.

Managed service providers.

Consultants.

Business partners.

Each introduces operational dependency.

Operational Evidence should demonstrate that governance extends beyond organizational boundaries.

Examples include:

Vendor security assessments.

. . .

Contractual security requirements.

Business continuity reviews.

Service-level monitoring.

Periodic reassessments.

Performance reviews.

Corrective action tracking.

Leadership cannot outsource governance.

Operational Evidence demonstrates that external relationships remain subject to organizational oversight.

Business Continuity and Resilience

Cybersecurity governance is ultimately about organizational resilience.

Resilience is not demonstrated through written plans alone.

. . .

It is demonstrated through preparation.

Organizations should produce Operational Evidence showing:

Business continuity planning.

Disaster recovery testing.

Backup validation.

Recovery exercises.

Critical system restoration testing.

Dependency analysis.

Operational recovery metrics.

Recovery improvement initiatives.

These activities provide confidence that governance decisions support organizational continuity.

. . .

Without testing, continuity plans remain assumptions.

Operational Evidence transforms assumptions into demonstrated capability.

Operational Metrics That Matter

Organizations collect enormous quantities of cybersecurity metrics.

Many provide little governance value.

Operational Evidence should focus on metrics that demonstrate organizational capability rather than technical activity.

Examples include:

Incident response time.

Mean time to recover.

Completion of corrective actions.

Training participation.

. . .

Tabletop exercise frequency.

Business continuity testing success.

Critical vendor assessments completed.

Policy implementation progress.

Management review completion.

Operational metrics should answer a simple question:

Is governance producing measurable operational improvement?

If the answer is yes, the metrics have value.

If not, leadership should reconsider what it is measuring.

Operational Evidence Ages Quickly

Unlike Governance Evidence, Operational Evidence has a relatively short useful life.

. . .

An incident response exercise conducted four years ago provides limited assurance today.

Training completed several years ago says little about current readiness.

A disaster recovery test performed on obsolete infrastructure no longer demonstrates resilience.

Operational Evidence must remain current.

This does not diminish the value of historical records.

Historical evidence demonstrates continuous improvement.

Current evidence demonstrates present capability.

Together, they tell a complete governance story.

Common Operational Evidence Gaps

Organizations frequently discover recurring weaknesses when evaluating Operational Evidence.

Incident response plans exist but have never been exercised.

. . .

Tabletop exercises occur without documenting lessons learned.

Corrective actions remain open indefinitely.

Security awareness training emphasizes completion rather than effectiveness.

Business continuity plans receive minimal testing.

Vendor oversight ends after contract signing.

Operational metrics measure activity rather than resilience.

Policies are implemented inconsistently across business units.

None of these gaps necessarily indicate operational failure.

They indicate opportunities to strengthen governance through more disciplined execution.

Building Better Operational Evidence

Organizations strengthen Operational Evidence by making governance visible in everyday operations.

. . .

Exercises become regular rather than occasional.

Corrective actions remain visible until verified.

Training evolves continuously.

Recovery capabilities are tested under realistic conditions.

Operational reviews become part of normal management practice.

Most importantly, leadership consistently asks:

How do we know this governance decision was actually implemented?

That question transforms governance from policy into practice.

Operational Evidence Completes the Connection

Operational Evidence occupies a unique position within the Defensible Evidence Framework™.

. . .

Governance Evidence establishes authority.

Risk Evidence informs decisions.

Oversight Evidence monitors leadership.

Operational Evidence demonstrates implementation.

It is where governance becomes visible.

Where strategy becomes execution.

Where leadership expectations become organizational behavior.

Without Operational Evidence, governance remains theoretical.

With it, organizations can demonstrate that cybersecurity governance influenced how people worked, how decisions were made, how incidents were managed, and how resilience improved over time.

That evidence becomes particularly powerful because it shows something investigators, regulators, insurers, and stakeholders all seek to understand:

. . .

Leadership did not merely establish cybersecurity governance.

Leadership made it operational.

Chapter 9
Assurance Evidence

Trust is essential.

Verification is indispensable.

Every effective governance system is built upon both.

Organizations trust executives to lead.

Executives trust managers to execute.

Managers trust technical teams to implement controls.

Customers trust organizations to protect their information.

Boards trust management's reporting.

Yet governance cannot depend on trust alone.

. . .

It requires assurance.

Assurance is the discipline of independently determining whether governance is functioning as leadership intends. It answers a question that every governing body should periodically ask:

How do we know?

How do we know the reported controls are actually operating?

How do we know policies are being followed?

How do we know risks are being managed?

How do we know corrective actions were completed?

How do we know the organization is becoming more resilient?

Assurance Evidence provides those answers.

Assurance Is Independent Confidence

One of the defining characteristics of assurance is independence.

. . .

Management evaluates its own performance every day.

That is appropriate.

Boards, however, require confidence that extends beyond self-assessment.

This confidence comes from independent validation.

Internal audit.

External assessments.

Penetration testing.

Control reviews.

Compliance assessments.

Third-party audits.

Governance evaluations.

. . .

These activities provide leadership with objective information that supports informed oversight.

Assurance is not about distrust.

It is about confidence supported by evidence.

Management Cannot Assure Itself

Organizations occasionally blur the distinction between reporting and assurance.

Management prepares a cybersecurity dashboard.

Management reports that critical controls are functioning.

Management concludes that governance is effective.

While these reports are valuable, they do not constitute assurance.

They remain management's assessment of management.

Assurance requires a degree of independence from the activities being evaluated.

. . .

This independence strengthens governance by providing leadership with multiple perspectives.

One perspective comes from management.

Another comes from assurance.

Good governance values both.

Assurance Completes the Governance Cycle

The five evidence domains of the Defensible Evidence Framework™ follow a natural progression.

Governance establishes direction.

Risk informs decisions.

Oversight monitors progress.

Operations execute leadership's direction.

Assurance validates the results.

. . .

Without assurance, leadership is left to assume that governance is functioning as intended.

With assurance, leadership gains objective evidence supporting that conclusion.

This distinction is more significant than it first appears.

Assumptions cannot be audited.

Evidence can.

Internal Audit's Unique Role

Among all assurance activities, internal audit occupies a particularly important position.

Unlike operational management, internal audit reports independently to the Board or an appropriate Board committee.

Its purpose is not to operate cybersecurity.

Its purpose is to evaluate whether governance, risk management, and internal controls are functioning effectively.

Strong Assurance Evidence often includes:

. . .

Internal audit reports.

Management responses.

Corrective action plans.

Follow-up reviews.

Closure verification.

Trend analysis.

Board presentations.

Collectively, these artifacts demonstrate that leadership actively evaluates governance rather than assuming its effectiveness.

External Assessments Add Credibility

Independent external assessments often provide additional assurance beyond internal governance activities.

These assessments may include:

. . .

Cybersecurity maturity assessments.

Independent governance reviews.

Penetration testing.

Red team exercises.

Regulatory examinations.

Certification audits.

Third-party compliance reviews.

Their value extends beyond technical findings.

External assessors frequently identify governance weaknesses that become invisible to organizations because of familiarity or organizational bias.

An experienced outside reviewer often asks questions leadership no longer thinks to ask.

That perspective strengthens governance.

Penetration Testing Is Governance Evidence

Penetration testing is frequently viewed as a purely technical activity.

It is much more than that.

From a governance perspective, penetration testing demonstrates that leadership sought objective evaluation of the organization's defenses.

Equally important is what happens after the testing concludes.

Were findings prioritized?

Were executive leaders informed?

Were material issues escalated?

Were remediation plans developed?

Did leadership verify completion?

The penetration test itself is valuable.

. . .

The governance response to its findings is even more valuable.

That response becomes Assurance Evidence.

Corrective Actions Are Where Assurance Becomes Real

Finding weaknesses is easy.

Correcting them is governance.

Organizations often accumulate assessment reports while failing to resolve recurring deficiencies.

The result is an expanding archive of identified problems accompanied by little evidence of improvement.

Strong Assurance Evidence demonstrates the complete lifecycle of corrective action.

Finding documented.

Management response approved.

Responsibility assigned.

. . .

Completion date established.

Progress monitored.

Implementation verified.

Closure documented.

This lifecycle transforms assurance from observation into organizational improvement.

Without corrective action, assessments become historical records.

With corrective action, they become governance instruments.

Assurance Is Continuous

Many organizations treat assurance as an annual event.

Annual audits.

Annual assessments.

Annual compliance reviews.

. . .

Annual certifications.

Cyber risk does not evolve annually.

Neither should assurance.

Effective governance establishes a continuous assurance program.

Routine internal reviews.

Periodic external assessments.

Ongoing vulnerability validation.

Recurring maturity evaluations.

Continuous monitoring of corrective actions.

Independent verification of significant governance activities.

The objective is not to audit everything constantly.

. . .

The objective is to provide leadership with continuous confidence that governance remains effective despite changing risks.

Assurance Supports Board Confidence

Board members are expected to exercise informed oversight.

That expectation creates a practical challenge.

Directors cannot personally verify every cybersecurity control.

They cannot inspect every server.

Review every configuration.

Validate every operational process.

Nor should they.

Their confidence comes from governance.

Governance depends upon assurance.

Assurance Evidence therefore enables Boards to fulfill their fiduciary responsibilities without becoming operational managers.

. . .

Independent validation provides confidence that leadership's decisions are producing intended results.

Confidence supported by evidence is far stronger than confidence based upon assumptions.

Assurance Is Not About Finding Fault

Some organizations approach audits and assessments defensively.

Findings are viewed as criticism.

Assessments become exercises in minimizing deficiencies.

This mindset weakens governance.

The purpose of assurance is not to assign blame.

It is to improve organizational resilience.

Organizations with mature governance welcome independent assessment because it identifies opportunities before adversaries, regulators, or auditors discover them under less favorable circumstances.

. . .

Every finding represents an opportunity to strengthen governance.

Every verified improvement becomes evidence that leadership learns and adapts.

That is precisely what reasonable governance should demonstrate.

Common Assurance Evidence Gaps

Organizations reviewing their Assurance Evidence frequently encounter recurring weaknesses.

Assessments occur irregularly.

Internal audit has limited cybersecurity expertise.

External assessments focus exclusively on technical controls.

Corrective actions remain open indefinitely.

Closure occurs without independent verification.

Findings are repeated year after year.

. . .

Board reporting emphasizes completed audits rather than governance improvements.

Independent assessments are performed only to satisfy regulatory requirements.

These gaps rarely indicate a lack of concern.

More often, they indicate that assurance has been treated as compliance rather than governance.

Building Better Assurance Evidence

Organizations strengthen Assurance Evidence by integrating independent validation into normal governance activities.

Assessments become expected rather than exceptional.

Findings become opportunities rather than embarrassments.

Corrective actions receive sustained executive attention.

Boards periodically request independent perspectives.

Management welcomes constructive challenge.

. . .

Verification becomes part of organizational culture.

Over time, assurance evolves from an audit function into a strategic governance capability.

That evolution significantly strengthens defensibility.

Assurance Creates Confidence

The Defensible Evidence Framework™ begins with governance and concludes with assurance for a reason.

Leadership establishes governance.

Management implements it.

Oversight monitors it.

Operations execute it.

Assurance confirms it.

When these five evidence domains operate together, they create something every organization seeks but few can confidently demonstrate:

Reasonable cybersecurity governance supported by objective evidence.

That evidence tells a compelling story.

Not because it claims perfection.

Not because it promises immunity from cyber incidents.

But because it demonstrates that leadership consistently sought to verify—not merely assume—that governance was functioning as intended.

And in the aftermath of a significant cyber event, that may be one of the strongest forms of evidence an organization can produce.

Chapter 10
How the Five Domains Work Together

The five evidence domains of the Defensible Evidence Framework™ are designed to function as an integrated governance system.

Individually, each domain answers an important question.

Together, they answer the only question that ultimately matters:

Can the organization demonstrate that it exercised reasonable cybersecurity governance?

That distinction is significant.

An organization may have excellent policies but poor oversight.

It may perform comprehensive risk assessments but never verify that corrective actions were completed.

. . .

It may conduct regular tabletop exercises but fail to communicate material risks to the Board.

Each of these organizations possesses evidence.

None possesses complete governance evidence.

The strength of the Defensible Evidence Framework™ lies not in any individual domain, but in how the five domains reinforce one another to create a complete and defensible governance narrative.

Governance Is a System

Organizations often approach cybersecurity governance as a collection of unrelated activities.

Policies are written by one group.

Risk assessments are performed by another.

Internal audit operates independently.

Executive reporting follows its own schedule.

Incident response planning belongs to technical teams.

. . .

Each activity may be well executed.

Yet if they remain disconnected, leadership struggles to demonstrate that governance functioned as a coordinated system.

Effective governance resembles an ecosystem rather than a checklist.

Every decision influences another.

Every governance activity creates evidence that supports the next.

The five evidence domains describe that ecosystem.

Governance establishes authority.

Risk identifies what matters.

Oversight monitors progress.

Operations execute leadership's direction.

Assurance validates the results.

. . .

Each domain depends upon the others.

Following a Single Governance Decision

Consider a simple example.

A quarterly enterprise cyber risk assessment identifies an increasing risk associated with third-party software suppliers.

Leadership determines that stronger vendor oversight is necessary.

That single governance decision produces evidence across every domain.

Governance Evidence

The Board directs management to strengthen third-party governance.

Committee minutes document the discussion.

A revised vendor risk policy is approved.

Responsibilities are assigned.

Authority is clarified.

Risk Evidence

The enterprise risk register is updated.

Third-party risk receives a higher priority.

Business impact analysis is revised.

Management evaluates mitigation options.

Residual risks are documented.

Oversight Evidence

The Board requests quarterly vendor risk reporting.

Management provides progress updates.

Outstanding issues remain on committee agendas until resolved.

Follow-up actions are tracked.

Operational Evidence

Vendor assessments are expanded.

. . .

Existing contracts are reviewed.

Critical suppliers undergo additional evaluation.

Business continuity plans are updated.

Vendor monitoring procedures are implemented.

Assurance Evidence

Internal audit validates implementation.

Independent assessors review vendor governance.

Corrective actions are verified.

Leadership receives objective assurance that governance changes became operational reality.

One governance decision.

Five evidence domains.

A complete governance story.

The Evidence Chain

The framework is intentionally sequential.

When these activities occur consistently, the organization develops what might be called an **evidence chain**.

Like any chain, its strength depends upon every link.

Missing Governance Evidence weakens accountability.

Weak Risk Evidence undermines decision-making.

Poor Oversight Evidence obscures leadership engagement.

. . .

Limited Operational Evidence raises questions about implementation.

Insufficient Assurance Evidence reduces confidence.

The objective is not perfection within each domain.

It is continuity across all five.

Governance Becomes Traceable

One of the framework's greatest advantages is traceability.

Investigators rarely examine governance through isolated documents.

They reconstruct events.

They connect decisions.

They establish chronology.

The Defensible Evidence Framework™ naturally supports this process.

Consider the following sequence.

. . .

A Board receives a quarterly cybersecurity briefing.

The briefing identifies increasing ransomware risk.

Leadership requests additional analysis.

Management recommends enhanced backup capabilities.

Funding is approved.

Implementation begins.

A tabletop exercise validates recovery procedures.

Internal audit confirms successful implementation.

Months later, the organization experiences a ransomware attack.

Because governance was documented throughout the process, leadership can demonstrate:

The risk was identified.

. . .

The Board discussed it.

Resources were allocated.

Management implemented improvements.

Recovery capabilities were tested.

Independent validation occurred.

Whether or not the attack succeeded becomes only part of the story.

Governance becomes equally visible.

The Narrative Matters

Every cyber incident eventually creates a narrative.

Sometimes the organization tells it.

Sometimes regulators do.

Sometimes attorneys do.

. . .

Sometimes journalists do.

Evidence determines which narrative prevails.

Without an organized governance record, the narrative often becomes fragmented.

Policies appear disconnected from risk assessments.

Board minutes seem unrelated to operational decisions.

Audit findings exist without documented corrective actions.

Investigators fill the gaps with assumptions.

The Defensible Evidence Framework™ organizes evidence into a coherent narrative.

Leadership established governance.

Leadership understood risk.

. . .

Leadership exercised oversight.

Leadership directed action.

Leadership verified results.

That sequence demonstrates far more than documentation.

It demonstrates disciplined governance.

Finding Weak Links

Organizations frequently discover that one evidence domain has matured significantly while another has been neglected.

For example:

Governance structures may be well established, but risk acceptance remains undocumented.

Operational execution may be excellent, but Board oversight lacks meaningful discussion.

Internal audit may identify recurring deficiencies, yet corrective actions remain unresolved.

. . .

The framework makes these weaknesses visible.

Rather than evaluating governance as a vague concept, organizations can identify specific evidence domains requiring improvement.

This enables governance maturity to progress systematically rather than reactively.

The Framework Is Not Linear

Although the five domains are presented sequentially, governance itself is continuous.

Assurance findings often identify new risks.

New risks prompt governance decisions.

Governance decisions create additional operational activities.

Operational improvements generate new assurance opportunities.

The cycle repeats.

Each iteration strengthens the governance evidence repository.

. . .

Each cycle increases organizational maturity.

Each improvement enhances defensibility.

The framework therefore functions less like a straight line and more like a continuous governance loop.

Organizations do not complete governance.

They continuously improve it.

Evidence Should Never Be Manufactured

One caution deserves emphasis.

Organizations should never attempt to create evidence simply because evidence is expected.

Manufactured evidence is easy to recognize.

Documents created after the fact.

Meeting minutes reconstructed months later.

Policies approved without meaningful discussion.

. . .

Risk acceptance forms signed solely to satisfy auditors.

These activities weaken credibility.

Evidence derives its value from authenticity.

The strongest Governance Evidence is created naturally while governance occurs—not reconstructed after questions arise.

That principle appears repeatedly throughout this book because it defines the difference between documentation and defensibility.

Seeing the Whole Picture

Many governance reviews focus on isolated artifacts.

A policy.

An audit report.

A risk assessment.

A committee charter.

. . .

Each document provides value.

None tells the entire story.

The Defensible Evidence Framework™ encourages leadership to view governance as an interconnected system of evidence.

When all five domains function together, the organization can answer questions that no single document could address.

How was governance established?

How were risks evaluated?

How did leadership exercise oversight?

How were decisions implemented?

How did leadership verify success?

Collectively, those answers describe reasonable cybersecurity governance.

Preparing for Implementation

Understanding the framework is only the first step.

The next challenge is implementation.

Where should organizations begin?

How should governance evidence be organized?

Who owns it?

How should it be maintained?

How can leadership determine whether important evidence is missing?

The next part of this book answers those questions by moving from theory to practice.

It begins with one of the most important—but frequently overlooked—components of defensible governance:

the Governance Evidence Repository.

Part III
Building Evidence

Chapter 11
Building the Governance Evidence Repository

Every organization possesses governance evidence.

The question is whether anyone can find it.

One of the most common discoveries made during governance assessments is that evidence already exists—but it is scattered across the organization. Board minutes reside in one system. Policies are maintained somewhere else. Risk registers belong to another application. Internal audit reports are archived separately. Tabletop exercise documentation lives in a project folder. Vendor assessments are maintained by Procurement. Incident reports remain with the security team.

Individually, these documents have value.

Collectively, they tell the story of governance.

Unless they can be found.

. . .

That is the purpose of the Governance Evidence Repository.

It is not another document management system.

It is not another compliance database.

It is the organizational memory of cybersecurity governance.

Governance Is Only as Defensible as Its Evidence

Imagine receiving a regulatory inquiry requesting documentation demonstrating Board oversight of cybersecurity during the previous three years.

How long would it take to produce it?

Hours?

Weeks?

Would anyone know where everything is located?

Could leadership demonstrate:

- Board oversight?
- Executive reporting?
- Risk acceptance decisions?
- Policy approvals?
- Audit findings?
- Corrective actions?
- Tabletop exercises?
- Independent assessments?

Or would the organization begin a frantic search through email archives, network drives, SharePoint sites, governance portals, and personal folders?

The speed with which evidence can be located often says as much about governance maturity as the evidence itself.

What Is a Governance Evidence Repository?

A Governance Evidence Repository is a structured collection of governance artifacts organized according to the five evidence domains of the Defensible Evidence Framework™.

Its purpose is straightforward:

To preserve governance evidence in a manner that demonstrates leadership oversight over time.

. . .

Think of it as a governance library rather than a filing cabinet.

Every document exists within context.

Every artifact contributes to a governance narrative.

Every governance activity leaves an evidentiary trail.

The repository simply preserves that trail.

Organize by Evidence Domain

Rather than organizing documents by department or document type, organize them according to the governance questions they answer.

For example:

Governance Evidence

- Governance policies
- Board resolutions
- Committee charters
- Organizational charts
- Delegations of authority
- Governance calendars
- Executive accountability assignments

Risk Evidence

- Enterprise risk assessments
- Risk registers
- Risk acceptance documentation
- Business impact analyses
- Third-party risk assessments
- Executive risk reports

Oversight Evidence

- Board minutes
- Committee minutes
- Executive dashboards
- Corrective action tracking
- Governance reports
- Board education materials

Operational Evidence

- Incident response plans
- Tabletop exercise documentation
- Security awareness metrics
- Business continuity exercises
- Disaster recovery tests
- Vendor oversight documentation

Assurance Evidence

- Internal audit reports
- External assessments
- Penetration tests

- Maturity assessments
- Compliance reviews
- Corrective action verification

This organization allows reviewers to understand governance rather than merely locate documents.

Metadata Is Evidence

Documents alone are not enough.

Each governance artifact should include enough information to establish its authenticity and context.

Useful metadata includes:

- Title
- Owner
- Evidence domain
- Date created
- Date approved
- Review date
- Version
- Governing body
- Classification
- Retention schedule

Without metadata, documents become difficult to authenticate and nearly impossible to organize over long periods.

Preserve the Governance Timeline

Governance is chronological.

The repository should preserve that chronology.

An investigator should be able to reconstruct governance by asking questions such as:

What did leadership know?

When did leadership know it?

What decision was made?

Who made the decision?

What happened afterward?

Every governance artifact contributes another point on that timeline.

. . .

Over months and years, those points become the organization's governance history.

Avoid Creating Evidence After the Fact

Perhaps the greatest temptation following a cyber incident is reconstruction.

Missing meeting notes.

Undocumented decisions.

Unsigned approvals.

Unrecorded discussions.

Organizations often attempt to recreate these records after the event.

That approach rarely succeeds.

Reconstructed evidence lacks the credibility of contemporaneous documentation.

Dates become uncertain.

. . .

Memories differ.

Context disappears.

The strongest evidence is created naturally as governance occurs.

The repository preserves that evidence.

It should never manufacture it.

Ownership Matters

Someone must own the repository.

That does not necessarily mean owning every document.

Instead, ownership means ensuring that governance evidence is:

- Collected.
- Organized.
- Protected.
- Reviewed.
- Retained.
- Accessible when required.

Depending on the organization, this responsibility may belong to:

- Corporate Governance
- Enterprise Risk Management
- Internal Audit
- The Chief Information Security Officer
- The Corporate Secretary
- Legal
- Compliance

The specific owner matters less than ensuring clear accountability.

A repository without ownership eventually becomes incomplete.

Security Applies to Governance Evidence

Ironically, governance evidence often contains some of the organization's most sensitive information.

Board deliberations.

Security assessments.

Audit findings.

Risk acceptance decisions.

. . .

Incident investigations.

These documents deserve protection equal to—or greater than—the operational systems they describe.

The repository should therefore include appropriate:

- Access controls
- Encryption
- Version control
- Backup procedures
- Retention policies
- Legal hold procedures
- Audit logging

Protecting governance evidence is itself an act of governance.

Think Like an Investigator

One of the simplest ways to evaluate a repository is to imagine you are conducting an independent investigation.

Suppose someone asks:

> "Show me every governance decision related to ransomware during the last two years."

. . .

Could the repository answer?

Or consider:

"Show me every Board discussion regarding third-party cyber risk."

Could that evidence be produced?

If locating governance evidence requires dozens of emails and multiple phone calls, the repository is serving the organization poorly.

Good repositories answer governance questions—not merely store documents.

Technology Is Secondary

Organizations sometimes assume they need specialized governance software to build an effective repository.

They do not.

Many organizations successfully use existing enterprise content management platforms.

. . .

Others use governance portals.

Some leverage document management systems already deployed throughout the enterprise.

The technology is far less important than the structure.

A poorly organized repository in expensive software remains poorly organized.

A well-designed repository using existing tools often performs exceptionally well.

Governance—not technology—determines its effectiveness.

The Repository as Institutional Memory

Leadership changes.

Board members rotate.

Executives retire.

Organizations merge.

. . .

Structures evolve.

The Governance Evidence Repository preserves continuity through all of these transitions.

It becomes the institutional memory of cybersecurity governance.

New leaders inherit not only policies and procedures but also the documented history of how governance has evolved over time.

That continuity strengthens future decision-making.

It also strengthens defensibility.

Preparing for Assessment

Once the repository exists, organizations gain a new capability.

They can evaluate the completeness of their governance evidence before anyone else does.

Rather than waiting for an audit, regulatory examination, insurance renewal, or cyber incident, leadership can periodically review the repository and ask a simple question:

If someone asked us tomorrow to demonstrate reasonable cybersecurity governance, would our evidence tell the complete story?

That question leads naturally to the next chapter, where we introduce a practical method for measuring governance maturity through the **Governance Readiness Scorecard**.

Chapter 12
Evidence Lifecycle Management

Evidence has a lifecycle.

Like every other organizational asset, governance evidence is created, used, maintained, protected, and eventually retired. Yet many organizations devote considerable attention to creating governance evidence while giving little thought to how that evidence should be managed over time.

This oversight creates unnecessary risk.

An unsigned Board resolution has little evidentiary value.

A policy that cannot be traced to an approving authority loses credibility.

A risk register that has not been updated in three years tells an incomplete story.

. . .

Meeting minutes that disappear because of poor retention practices cannot support governance during litigation or regulatory review.

Effective governance requires more than creating evidence.

It requires managing evidence throughout its entire lifecycle.

That is the purpose of Evidence Lifecycle Management.

Governance Does Not End When a Document Is Written

Many organizations unconsciously treat documentation as the finish line.

A policy is approved.

A risk assessment is completed.

A committee meeting concludes.

An audit report is issued.

The work appears complete.

. . .

From an evidentiary perspective, however, the work has only begun.

Every governance artifact must continue to demonstrate authenticity, integrity, accessibility, and relevance throughout its useful life.

Evidence that cannot be trusted is no longer evidence.

It is simply information.

The Evidence Lifecycle

Every governance artifact moves through six distinct stages.

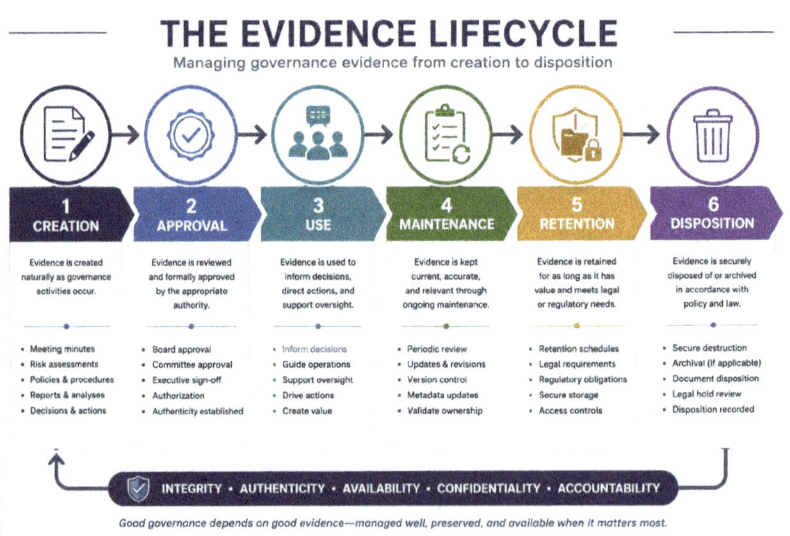

Each stage contributes to the credibility and defensibility of the evidence.

Weakness at any stage diminishes its value.

Stage One: Creation

Evidence should be created naturally as governance occurs.

Board meetings produce minutes.

Risk reviews produce updated risk registers.

Committee discussions produce decisions.

Policies produce approval records.

Tabletop exercises produce observations and corrective actions.

The key principle is contemporaneous documentation.

Governance should be documented while events are occurring—not reconstructed weeks or months later.

. . .

Contemporaneous evidence consistently carries greater credibility because it reflects decisions made in real time rather than memories reconstructed afterward.

Stage Two: Approval

Not every document becomes governance evidence simply because it exists.

Governance evidence requires authority.

Policies should demonstrate formal approval.

Committee charters should identify approving bodies.

Risk acceptance decisions should identify authorized decision-makers.

Board minutes should reflect formal acceptance according to organizational procedures.

Approval establishes authenticity.

Without it, documents may represent drafts, recommendations, or unofficial opinions rather than organizational decisions.

. . .

One of the simplest questions an investigator may ask is:

Who approved this?

Evidence should answer that question immediately.

Stage Three: Use

Governance evidence exists to support governance.

Policies should guide decisions.

Risk registers should inform executive discussions.

Dashboards should support Board oversight.

Audit reports should drive corrective action.

Evidence that is never referenced quickly loses practical value.

One indication of governance maturity is the degree to which existing evidence is routinely incorporated into leadership decision-making.

Documents should remain living governance instruments rather than historical artifacts.

Stage Four: Maintenance

Governance evolves.

Evidence must evolve with it.

Organizations change.

Technology changes.

Threats change.

Leadership changes.

Governance documents should reflect those changes.

Maintenance includes:

- Periodic review.
- Version updates.
- Approval of revisions.
- Removal of obsolete references.
- Validation of ownership.
- Updating metadata.

Maintenance demonstrates that governance remains active.

An outdated policy often reveals more about governance than the policy itself.

Current evidence demonstrates active leadership.

Obsolete evidence demonstrates historical leadership.

Both have value, but they serve different purposes.

Stage Five: Retention

Not every governance artifact should be retained forever.

Neither should important evidence be discarded prematurely.

Organizations should establish documented retention schedules consistent with:

Legal requirements.

Regulatory obligations.

Organizational policy.

. . .

Litigation considerations.

Operational needs.

Historical value.

Retention decisions should never occur accidentally.

Nor should they depend upon individual preferences.

Evidence supporting significant governance decisions may remain valuable for many years after the original decision was made.

Particularly following a cyber incident, historical governance records often become essential to reconstructing leadership's decision-making process.

Stage Six: Disposition

Eventually, governance evidence reaches the end of its useful life.

Disposition should occur deliberately.

Secure destruction.

. . .

Archival preservation.

Transfer to historical records.

Legal hold exceptions.

Every disposition action should itself be documented.

Organizations frequently maintain rigorous controls over evidence creation while exercising little control over evidence destruction.

Poor disposition practices create unnecessary legal and operational risk.

Governance should remain disciplined until the evidence lifecycle is complete.

Integrity Is Everything

The value of evidence depends upon its integrity.

Integrity means more than preventing unauthorized modification.

It also means preserving confidence.

. . .

Evidence should demonstrate:

Who created it.

When it was created.

Who modified it.

When modifications occurred.

Why revisions were necessary.

Who approved changes.

Version histories become particularly important for governance documents that evolve over time.

A policy with a documented revision history demonstrates continuous governance.

A policy with unexplained modifications raises questions.

. . .

Integrity creates credibility.

Credibility creates defensibility.

Metadata Protects Context

Evidence separated from its context often loses meaning.

Consider a committee meeting agenda.

Without additional information, investigators may struggle to determine:

Which committee met?

When?

Who attended?

Was a quorum present?

What decisions resulted?

Metadata preserves this context.

. . .

Every governance artifact should include sufficient metadata to establish its relationship to the larger governance process.

Useful metadata commonly includes:

- Title.
- Owner.
- Evidence domain.
- Date created.
- Date approved.
- Review frequency.
- Current version.
- Classification.
- Retention category.
- Related governance activities.

Metadata transforms isolated documents into connected governance evidence.

Legal Holds and Preservation

Following a significant cyber incident, organizations may become subject to litigation, regulatory investigation, insurance review, or law enforcement inquiry.

At that moment, evidence preservation becomes paramount.

Routine retention schedules may need to be suspended.

. . .

Disposition activities may stop.

Relevant governance artifacts should be preserved until legal obligations are satisfied.

Organizations should establish documented legal hold procedures before they become necessary.

Attempting to determine preservation requirements during a crisis introduces unnecessary uncertainty.

Good governance anticipates these situations.

Digital Preservation

Modern governance is overwhelmingly digital.

Board portals.

Email.

Collaboration platforms.

Document repositories.

. . .

Cloud storage.

Workflow systems.

Digital evidence introduces challenges beyond those associated with traditional paper records.

Organizations should consider:

Long-term file accessibility.

Migration between technology platforms.

Digital signatures.

Audit trails.

Backup integrity.

Access controls.

Protection against unauthorized alteration.

. . .

Preservation of electronic metadata.

Technology will continue to evolve.

Governance evidence must remain readable regardless of future technology changes.

Common Evidence Lifecycle Failures

Organizations evaluating Evidence Lifecycle Management frequently discover familiar weaknesses.

Policies without approval dates.

Risk registers with unclear ownership.

Meeting minutes lacking formal acceptance.

Documents stored in personal folders.

Inconsistent version control.

Missing retention schedules.

Evidence destroyed without documentation.

. . .

No legal hold procedures.

Obsolete governance artifacts remaining in active use.

None of these deficiencies necessarily indicate poor governance.

They indicate incomplete evidence management.

The distinction matters.

Strong governance deserves equally strong stewardship of its evidence.

Building a Sustainable Evidence Program

Evidence Lifecycle Management should not become an administrative burden.

The objective is not more documentation.

It is better stewardship.

Organizations should establish repeatable processes that ensure governance evidence remains:

- Authentic.
- Accessible.
- Current.
- Protected.
- Traceable.
- Reliable.

When these characteristics become routine, evidence retains its value throughout its lifecycle.

Evidence Is a Long-Term Asset

Cybersecurity technologies become obsolete.

Leadership changes.

Organizations restructure.

Regulations evolve.

Governance evidence often outlives all of them.

Years after a significant decision was made, that evidence may become the only reliable record explaining why leadership acted as it did.

. . .

For that reason, governance evidence should be treated as a strategic organizational asset—not merely as documentation created to satisfy audits or compliance requirements.

Evidence Lifecycle Management ensures that the value of governance does not diminish with time.

It preserves the integrity of leadership's decisions long after those decisions have been made.

Because the true value of governance evidence is not measured on the day it is created.

It is measured on the day someone asks to see it.

Chapter 13
Governance by Design

Organizations have embraced concepts such as security by design, privacy by design, and resilience by design because experience has shown that bolting critical capabilities onto a system after deployment is expensive, inconsistent, and often ineffective. The same principle applies to governance.

Governance should never be an afterthought.

Too many organizations develop governance only after a regulatory finding, a failed audit, a cyber incident, or an insurance requirement forces action. Policies are written in haste, committees are created without clear authority, reporting becomes reactive, and documentation is reconstructed from memory rather than produced as a natural by-product of leadership activities.

This approach creates the appearance of governance without delivering its benefits.

Governance by Design takes the opposite approach. It intentionally embeds governance into the way the organization operates so that accountability, oversight, decision-making, and evidence are created continuously rather than assembled after the fact.

Governance becomes part of the organizational architecture.

It is designed—not improvised.

Governance Is an Architectural Decision

Every organization has a governance architecture, whether intentionally designed or not.

Some organizations possess clearly defined authority, disciplined reporting, documented decisions, and well-understood accountability. Others rely on personalities, informal conversations, and institutional memory.

Both are governance models.

Only one is sustainable.

Just as enterprise architecture determines how technology components interact, governance architecture determines how leadership functions interact.

It answers questions such as:

- Who owns cyber risk?
- Who approves policy?
- Who accepts residual risk?
- What information reaches executive leadership?
- What information reaches the Board?
- How are decisions documented?
- How is effectiveness validated?

These questions should never require debate during a crisis.

They should already have been answered through design.

Governance Should Be Embedded in Business Processes

One of the most common misconceptions is that governance exists separately from operational work.

It does not.

Well-designed governance is embedded into existing business processes.

Risk assessments become part of strategic planning.

Policy reviews become part of annual governance cycles.

. . .

Cybersecurity reporting becomes part of regular Board meetings.

Vendor governance becomes part of procurement.

Exception management becomes part of operational decision-making.

Incident reviews become part of continuous improvement.

None of these activities require additional bureaucracy when governance has been intentionally designed. They become routine organizational behaviors that naturally produce oversight and accountability.

Governance succeeds when it becomes ordinary.

Designing for Accountability

Every governance design should begin with accountability.

Not responsibility.

Responsibility describes who performs work.

Accountability identifies who owns the outcome.

. . .

Those are rarely the same person.

Security teams implement controls.

Business leaders own business risk.

Compliance personnel evaluate adherence.

Internal audit validates effectiveness.

Executive leadership directs organizational priorities.

The Board exercises oversight.

When accountability is unclear, governance weakens regardless of how sophisticated the technology becomes.

Governance by Design deliberately eliminates ambiguity.

Every material decision should have a clearly identifiable owner.

Decision Rights Must Exist Before Decisions Are Needed

Organizations often discover governance gaps during moments of uncertainty.

Who decides whether systems should be disconnected?

Who authorizes public disclosure?

Who accepts operational disruption?

Who approves paying external incident responders?

Who informs regulators?

If these questions are being answered for the first time during an incident, governance has already failed.

Governance by Design establishes decision rights before they are needed.

Authority is documented.

Escalation paths are defined.

. . .

Approvals are understood.

Leadership practices them before a crisis demands their use.

Governance Should Produce Evidence Automatically

One of the defining characteristics of mature governance is that evidence is created naturally.

Meeting agendas.

Board packets.

Risk registers.

Policy approvals.

Exception requests.

Management reports.

Audit findings.

. . .

Executive decisions.

Committee minutes.

These are not administrative artifacts.

They are governance evidence.

Organizations that intentionally design governance processes rarely struggle to demonstrate oversight because evidence accumulates as leadership performs its normal responsibilities.

Organizations that fail to design governance often spend months attempting to reconstruct what occurred after regulators, auditors, insurers, or investigators begin asking questions.

Evidence should be created through governance—not recreated after governance.

Governance Must Survive Leadership Changes

Strong governance cannot depend upon individual personalities.

Executives retire.

. . .

Board members rotate.

Security leaders accept new positions.

Institutional knowledge disappears.

Governance by Design recognizes this reality by embedding governance into organizational structures rather than individual experience.

Policies define expectations.

Committee charters define authority.

Reporting standards define information flow.

Decision records preserve rationale.

Governance continues because the system survives the people who operate it.

This continuity is one of the hallmarks of organizational maturity.

Governance Is Never Finished

Design does not imply permanence.

Business models evolve.

Technology changes.

Threat actors adapt.

Regulations expand.

Boards change.

Governance architecture must evolve alongside the organization it serves.

This evolution should occur deliberately through periodic review rather than through emergency reaction.

The goal is not stability for its own sake.

The goal is disciplined adaptation.

. . .

Well-designed governance changes intentionally.

Poorly designed governance changes only after failure.

Governance Reality

The strongest governance systems are rarely the most visible. They operate quietly in the background, providing clear authority, disciplined oversight, consistent decision-making, and continuous evidence long before anyone asks whether governance existed.

Governance by Design ensures that when that question is eventually asked, the organization does not need to explain its governance—it can simply demonstrate it.

Part IV
Measuring Defensibility

Chapter 14
Governance Readiness Scorecard

What gets measured gets improved.

Organizations routinely measure cybersecurity performance through technical metrics.

Patch compliance.

Critical vulnerabilities.

Mean time to detect.

Mean time to recover.

Security awareness participation.

Penetration test results.

. . .

These metrics reveal how security operations are performing.

They reveal very little about governance.

The Governance Readiness Scorecard measures something different.

It evaluates an organization's ability to govern cybersecurity through demonstrable evidence.

Rather than asking, *"How secure are we?"*, it asks a far more important leadership question:

"How prepared are we to demonstrate that we exercised reasonable cybersecurity governance?"

That distinction changes the conversation from security performance to governance readiness.

Readiness Is Not the Same as Security

An organization can operate a sophisticated Security Operations Center, maintain excellent vulnerability management, and deploy world-class security technologies while remaining poorly prepared to demonstrate governance.

. . .

Likewise, an organization with fewer technical resources may possess disciplined governance, clear oversight, well-documented decisions, and a mature evidence repository.

Security measures protection.

Governance Readiness measures defensibility.

The Governance Readiness Scorecard therefore evaluates leadership —not technology.

Measuring Evidence Readiness

The assessment evaluates maturity across the five evidence domains introduced earlier in this book:

- Governance Evidence
- Risk Evidence
- Oversight Evidence
- Operational Evidence
- Assurance Evidence

Within each domain, leadership evaluates questions such as:

- Is governance clearly established?
- Is cyber risk consistently evaluated?
- Does executive leadership receive actionable information?
- Does the Board exercise meaningful oversight?

- Are governance decisions implemented operationally?
- Are governance activities independently validated?
- Can evidence be produced quickly when requested?

These questions measure governance capability rather than technical implementation.

Measuring Evidence Quality

Evidence exists on a spectrum.

Simply possessing documentation does not indicate governance maturity.

For example, three organizations may all possess cybersecurity policies.

The first created them five years ago and has never reviewed them.

The second reviews them annually.

The third reviews them annually, documents Board approval, tracks revisions, validates implementation, and preserves a complete history of changes.

All three have policies.

. . .

Only one demonstrates mature governance.

The Governance Readiness Scorecard therefore evaluates the quality of evidence rather than its mere existence.

High-quality evidence is:

- Authentic
- Current
- Traceable
- Complete
- Consistently maintained
- Clearly linked to governance activities

As evidence quality improves, organizational defensibility improves.

Level 1 — Reactive

Reactive organizations govern primarily in response to events.

Governance is informal.

Roles are unclear.

Documentation is fragmented.

. . .

Evidence is difficult to locate.

Leadership discussions often occur without producing durable records.

Cybersecurity governance is largely driven by incidents, audit findings, insurance requirements, or regulatory pressure.

Governance exists—but only after circumstances demand it.

At this stage, organizations typically struggle to demonstrate reasonable oversight because governance activities have not yet become disciplined or repeatable.

Level 2 — Developing

Developing organizations recognize the need for formal governance.

Governance structures begin to emerge.

Policies have been established.

Risk assessments occur periodically.

. . .

Evidence exists but remains inconsistent across departments.

Leadership has begun building governance, but processes still depend heavily on individual initiative rather than organizational discipline.

Governance is becoming intentional.

Evidence remains uneven.

Level 3 — Managed

Managed organizations operate with repeatable governance processes.

Governance activities are documented.

Board oversight occurs regularly.

Evidence is organized and maintained.

Operational activities align with governance expectations.

Leadership responsibilities are understood.

. . .

Decision-making follows established processes.

Rather than reacting to events, governance becomes part of normal organizational operations.

Level 4 — Measured

Measured organizations continuously evaluate the effectiveness of governance itself.

Governance performance is measured.

Evidence quality is monitored.

Independent assurance informs leadership decisions.

Corrective actions are tracked and verified.

Governance metrics extend beyond operational security measures and begin evaluating oversight effectiveness, decision quality, and organizational maturity.

Leadership no longer assumes governance is functioning.

It measures it.

Level 5 – Defensible

Defensible organizations represent the highest level of governance maturity.

Evidence forms a complete, integrated body of proof.

Leadership oversight is clearly demonstrated.

Cyber risks are governed deliberately.

Independent validation occurs routinely.

Governance remains resilient despite leadership changes, organizational growth, regulatory evolution, and technological change.

Evidence is no longer collected simply for audits.

It is created naturally through disciplined governance.

When regulators, insurers, investigators, auditors, or courts ask how cybersecurity was governed, the organization does not rely upon recollection.

. . .

It produces evidence.

This is the objective of the Defensible Evidence Framework™.

Not perfect cybersecurity.

Defensible governance.

The Goal Is Continuous Improvement

Organizations should not view the maturity ladder as a certification.

It is a roadmap.

Every organization begins somewhere.

Some are building governance for the first time.

Others possess mature governance but inconsistent evidence.

Still others have strong governance processes but lack integrated assurance.

The Governance Readiness Scorecard identifies where the organization stands today and provides direction for moving upward.

. . .

Progress matters more than perfection.

Looking Beyond the Score

The numerical score is not the most valuable outcome of the assessment.

The conversation is.

The assessment encourages Boards, executives, cybersecurity leaders, auditors, legal counsel, and risk professionals to discuss governance using a common vocabulary.

It reveals differing assumptions.

It identifies governance blind spots.

It highlights strengths.

It establishes priorities.

Most importantly, it helps leadership answer the question that inspired this framework.

The Monday Morning Test

The Governance Readiness Scorecard ultimately measures one thing:

How confidently can your organization answer the Monday Morning Question?

When the incident makes headlines...

When regulators begin asking questions...

When insurers request documentation...

When attorneys demand records...

When the Board seeks assurance...

Will your organization be able to demonstrate that it exercised reasonable cybersecurity governance?

Organizations operating at the **Reactive** level will struggle to answer.

. . .

Organizations at the **Developing** level will produce partial evidence.

Organizations at the **Managed** level will present a coherent governance record.

Organizations at the **Measured** level will demonstrate governance effectiveness through objective metrics.

Organizations at the **Defensible** level will produce an integrated body of evidence that tells the complete story of leadership oversight.

That is the purpose of the Governance Readiness Scorecard.

Not simply to measure governance maturity—

but to help organizations build governance capable of withstanding scrutiny long before anyone asks to see it.

GOVERNANCE MATURITY LADDER
A Continuum of Evidence Readiness
Cyber governance maturity is measured by the quality, consistency, and defensibility of evidence.
1
REACTIVE
• Governance is ad hoc and event-driven
• Roles are unclear
• Documentation is fragmented
• Evidence is difficult to locate
Governance exists in response to events.
2
DEVELOPING
• Governance structures are emerging
• Policies are established
• Risk assessments occur periodically
• Evidence is inconsistent across functions
Governance is recognized and being implemented.
3
MANAGED
• Governance processes are documented and repeatable
• Board oversight occurs regularly
• Evidence is organized and maintained
• Activities align with governance expectations
Governance is implemented consistently and monitored.
4
MEASURED
• Governance effectiveness is measured
• Evidence quality is monitored continuously
• Independent assurance informs decisions
• Improvements are proactive
Governance is measured, validated, and continuously improved.
5
DEFENSIBLE
• Evidence forms a complete, integrated body of proof
• Leadership oversight is clearly demonstrated
• Risks are governed deliberately
• Independent validation is routine
• Governance is resilient and sustainable
Governance is defensible, credible, and built to withstand scrutiny.
THE GOAL: BE PREPARED TO ANSWER THE MONDAY MORNING QUESTION.
When the incident makes headlines, can we demonstrate that we exercised reasonable cybersecurity governance?
DOCUMENTED
CONSISTENT
MEASURABLE
AUDITABLE
DEFENSIBLE

Chapter 15
Governance Maturity

Governance is not a destination.

It is a discipline.

Organizations often speak about becoming "mature," as though governance reaches a point where no further improvement is necessary. In reality, governance maturity is not a final achievement. It is the organization's capacity to continuously govern changing risks through disciplined leadership, informed decision-making, and defensible evidence.

Technology changes.

Threat actors change.

Business priorities change.

Regulations change.

. . .

Leadership changes.

Consequently, governance must continue changing as well.

The organizations that thrive are not those with the most documentation.

They are the ones whose governance continually evolves.

Maturity Is More Than Process

Many maturity models emphasize process.

Processes matter.

They create consistency.

They improve repeatability.

They reduce uncertainty.

But mature governance extends beyond process.

True governance maturity demonstrates that leadership consistently exercises informed judgment while creating evidence that explains its decisions.

In other words, maturity is measured not simply by what an organization does.

It is measured by how well leadership governs.

The Characteristics of Mature Governance

Organizations with mature cybersecurity governance tend to share several characteristics.

Leadership understands that cybersecurity is an enterprise risk rather than an information technology problem.

Boards regularly discuss cyber risk as part of normal governance activities.

Executive leadership receives meaningful information that supports decision-making.

Roles and accountability are clearly defined.

Risk acceptance occurs deliberately.

. . .

Independent assurance validates governance.

Evidence is organized, protected, and readily accessible.

Most importantly, governance occurs continuously rather than episodically.

These organizations do not wait for audits to improve governance.

They improve governance because it improves leadership.

Governance Becomes Organizational Culture

One of the defining characteristics of mature organizations is that governance no longer feels like additional work.

It becomes part of organizational culture.

Board members instinctively ask governance questions.

Executives routinely document significant decisions.

Managers understand escalation expectations.

Risk discussions become part of strategic planning.

. . .

Corrective actions are followed until completion.

Evidence accumulates naturally because governance has become embedded in normal operations.

Culture is perhaps the highest expression of governance maturity.

When governance becomes habit, evidence follows naturally.

Leadership Drives Maturity

Governance maturity cannot be delegated.

Security professionals contribute expertise.

Risk professionals facilitate assessments.

Auditors provide assurance.

Legal counsel advises on obligations.

Compliance teams monitor requirements.

. . .

None of these functions can create governance maturity independently.

Only leadership can do that.

Boards establish expectations.

Executives set priorities.

Management reinforces accountability.

Every improvement in governance maturity ultimately begins with leadership commitment.

Technology enables governance.

Leadership defines it.

Maturity Is Demonstrated During Uncertainty

Governance is easy when conditions are favorable.

Budgets are adequate.

Operations are stable.

. . .

Threat activity is low.

Stakeholders are satisfied.

The true measure of governance maturity emerges during uncertainty.

A major cyber incident.

A regulatory investigation.

A ransomware attack.

A critical vendor failure.

A significant data breach.

Organizations with immature governance often improvise.

Authority becomes unclear.

Reporting becomes inconsistent.

. . .

Documentation becomes fragmented.

Decision-making slows.

Organizations with mature governance behave differently.

Decision authority is already established.

Escalation procedures function as expected.

Leadership receives accurate information.

Governance continues despite operational disruption.

Evidence is created as decisions are made.

Crises do not create governance maturity.

They reveal it.

Maturity Requires Continuous Learning

Every governance activity creates an opportunity to improve.

Board discussions reveal better questions.

Risk assessments expose changing priorities.

Tabletop exercises identify weaknesses.

Audits uncover process gaps.

Incident reviews highlight opportunities.

Mature organizations systematically convert these lessons into governance improvements.

Policies are updated.

Reporting evolves.

Training improves.

Committee charters are refined.

. . .

Risk methodologies mature.

Corrective actions are completed.

Governance itself becomes a learning system.

This continual refinement distinguishes mature governance from static governance.

Beware of Governance Complacency

Perhaps the greatest threat to governance maturity is success itself.

Organizations that avoid significant cyber incidents often assume their governance is effective.

That assumption can be dangerous.

Absence of incidents does not prove governance maturity.

It may simply reflect favorable circumstances.

Mature organizations continually question themselves.

. . .

What assumptions are we making?

What evidence is missing?

What has changed?

Where are our governance blind spots?

What would an external reviewer conclude?

These questions prevent complacency.

Good governance remains curious.

Measuring Progress

Governance maturity should be evaluated over time.

Not simply through periodic assessments.

Leadership should observe meaningful trends.

Are Board discussions becoming more strategic?

. . .

Is evidence becoming easier to locate?

Are corrective actions closing more quickly?

Is risk reporting supporting better decisions?

Are assurance findings decreasing?

Is governance becoming more consistent?

Progress should be measured through organizational capability rather than numerical scores alone.

The objective is continual improvement.

The Ultimate Measure of Maturity

The highest level of governance maturity is not demonstrated by certifications.

Nor by audit results.

Nor by regulatory compliance.

It is demonstrated when governance becomes self-sustaining.

. . .

Leadership naturally asks the right questions.

Decision-making follows established principles.

Evidence is created without special effort.

Assurance continuously validates performance.

Improvement becomes routine.

At that point, governance ceases to be a project.

It becomes part of the organization's identity.

Governance as a Competitive Advantage

Historically, organizations viewed cybersecurity governance as a defensive necessity.

Increasingly, it has become a strategic advantage.

Organizations with mature governance often experience:

. . .

Greater Board confidence.

Improved executive decision-making.

Stronger insurer relationships.

More productive regulatory interactions.

Higher stakeholder trust.

Faster incident recovery.

Greater organizational resilience.

Governance maturity therefore creates value beyond risk reduction.

It strengthens leadership itself.

The Defensible Organization

The ultimate purpose of governance maturity is not merely to withstand cyber attacks.

It is to create organizations capable of demonstrating disciplined leadership under scrutiny.

. . .

A defensible organization does not claim to have eliminated cyber risk.

It demonstrates that leadership understood the risks, governed them responsibly, monitored them continuously, adapted as circumstances changed, and preserved the evidence necessary to prove those actions.

That is the highest expression of governance maturity.

Not the absence of incidents.

Not perfect compliance.

Not flawless security.

But leadership that can withstand scrutiny because its governance is visible, intentional, repeatable, and supported by evidence.

That is the destination toward which every chapter in this book has been leading.

It is the defining characteristic of an organization whose governance is truly defensible.

Chapter 16
Evidence Gap Analysis

The absence of evidence does not always mean governance failed.

It may simply mean governance cannot be demonstrated.

That distinction lies at the heart of the Defensible Evidence Framework™.

Organizations often assume they understand the maturity of their cybersecurity governance because they have policies, conduct risk assessments, perform audits, and present reports to executive leadership. Yet when asked to produce evidence supporting those activities, they frequently discover missing documentation, inconsistent practices, unclear ownership, or incomplete governance records.

These are evidence gaps.

An Evidence Gap Analysis is the systematic process of identifying the difference between the governance evidence that should exist and the evidence that actually exists.

. . .

It is one of the most valuable exercises an organization can perform because it shifts governance from assumption to demonstration.

Every Governance Activity Leaves Evidence

Throughout this book, one principle has remained constant.

Good governance naturally produces evidence.

Board meetings produce minutes.

Risk reviews produce updated risk registers.

Policy approvals produce documented authority.

Tabletop exercises produce lessons learned.

Audits produce findings and corrective actions.

If governance activities occurred, evidence should exist.

When expected evidence is missing, leadership should ask a simple question:

Did the governance activity fail to occur, or did we fail to preserve the evidence?

The answer determines the organization's next step.

What Is an Evidence Gap?

An evidence gap exists whenever governance cannot be adequately demonstrated.

Sometimes the activity occurred but was never documented.

Sometimes documentation exists but cannot be located.

Sometimes ownership is unclear.

Sometimes evidence exists but lacks sufficient detail to demonstrate meaningful governance.

For example:

The Board discussed cybersecurity, but the minutes record only that "cybersecurity was reviewed."

. . .

A risk was accepted, but no documentation identifies who accepted it.

An audit finding was resolved, but no verification demonstrates completion.

A tabletop exercise occurred, but no after-action report was produced.

In each case, governance may have occurred.

The evidence does not prove it.

Evidence Gaps Are Governance Risks

Organizations often focus exclusively on operational cyber risks.

Unpatched systems.

Weak passwords.

Third-party vulnerabilities.

Cloud misconfigurations.

. . .

These are important.

Evidence gaps are governance risks.

Following a cyber incident, missing evidence creates uncertainty.

Uncertainty increases scrutiny.

Scrutiny increases organizational exposure.

Leadership cannot explain decisions it cannot document.

Evidence gaps therefore deserve the same disciplined attention applied to operational cybersecurity risks.

Conducting an Evidence Gap Analysis

A successful analysis begins with a straightforward question.

What evidence should exist if governance is functioning as intended?

The answer comes directly from the five evidence domains.

Governance Evidence

Can the organization produce:

- Current governance policies?
- Committee charters?
- Board-approved governance structures?
- Delegations of authority?
- Governance calendars?

Risk Evidence

Can leadership produce:

- Current risk registers?
- Business impact analyses?
- Risk acceptance documentation?
- Executive risk reports?
- Third-party risk assessments?

Oversight Evidence

Can the organization demonstrate:

- Board discussions?
- Committee oversight?
- Executive reporting?
- Corrective action follow-up?
- Governance dashboards?

Operational Evidence

Can leadership locate:

- Incident response documentation?
- Tabletop exercise reports?
- Business continuity testing?
- Vendor oversight records?
- Training metrics?

Assurance Evidence

Can the organization produce:

- Internal audit reports?
- External assessments?
- Penetration testing results?
- Corrective action verification?
- Independent governance reviews?

Every missing artifact represents an opportunity to strengthen governance.

Assessing Evidence Quality

Finding evidence is only the first step.

The next question is whether that evidence is defensible.

Consider two Board minutes.

One states:

"Cybersecurity was discussed."

The other states:

"The Board reviewed the quarterly cyber risk report, discussed increasing ransomware exposure affecting manufacturing operations, approved additional funding for immutable backups, and directed management to report implementation progress during the next quarterly meeting."

Both are meeting minutes.

Only one demonstrates governance.

Evidence Gap Analysis therefore evaluates quality as carefully as quantity.

Useful evidence should be:

- Current.
- Authentic.
- Complete.

- Traceable.
- Relevant.
- Approved.
- Accessible.

Evidence that lacks these characteristics often provides little governance value.

Looking for Patterns

Individual gaps matter.

Patterns matter more.

Organizations frequently discover recurring themes.

Documentation exists but ownership is inconsistent.

Policies are current, but approval records are missing.

Risk assessments occur regularly, but risk acceptance remains undocumented.

Board reporting is consistent, but follow-up actions disappear after meetings.

. . .

Audits identify recurring deficiencies that remain unresolved.

Patterns reveal systemic governance weaknesses.

Those weaknesses deserve executive attention because they affect multiple evidence domains simultaneously.

Prioritizing Gaps

Not every gap carries the same level of risk.

Organizations should prioritize improvements according to governance impact.

High-priority gaps generally involve:

Undefined governance authority.

Undocumented Board oversight.

Missing risk acceptance.

Absent incident governance.

Incomplete assurance activities.

. . .

Evidence supporting significant regulatory obligations.

Moderate-priority gaps often involve consistency, organization, or metadata improvements.

Lower-priority gaps typically involve administrative refinements that improve efficiency without significantly affecting defensibility.

Prioritization ensures governance improvements occur where they provide the greatest value.

Closing the Gap

Evidence gaps should never be addressed by creating documents solely for appearance.

The objective is to improve governance—not paperwork.

If Board oversight is weak, improve oversight.

If risk acceptance is inconsistent, formalize the decision process.

If corrective actions lack follow-up, strengthen accountability.

. . .

Improved governance naturally produces improved evidence.

Attempting to solve evidence gaps through documentation alone merely creates better records of inadequate governance.

The evidence must reflect reality.

Evidence Gap Analysis Is Continuous

Organizations should not perform an Evidence Gap Analysis only after a cyber incident.

Nor should it occur only before an audit.

Evidence gaps evolve.

Leadership changes.

Business priorities shift.

Technology platforms are replaced.

Regulatory expectations mature.

. . .

Consequently, Evidence Gap Analysis should become part of the organization's governance cycle.

Many organizations conduct it annually.

Others incorporate it into internal audit activities or governance committee reviews.

The specific schedule matters less than the discipline.

Governance improves through continual evaluation.

The Governance Conversation

One of the greatest benefits of an Evidence Gap Analysis is the discussion it creates.

Boards gain visibility into governance strengths and weaknesses.

Executive leadership develops a clearer understanding of organizational accountability.

Security leaders learn where reporting can improve.

Internal auditors identify opportunities for future assurance work.

. . .

Legal and compliance teams better understand evidentiary risk.

The analysis becomes more than an assessment.

It becomes a catalyst for governance maturity.

From Gaps to Defensibility

Every organization has evidence gaps.

Even mature organizations continuously discover opportunities to improve.

The objective is not to eliminate every gap.

It is to ensure that significant governance activities consistently produce credible, organized, and accessible evidence.

Over time, those improvements accumulate.

Evidence becomes more complete.

Governance becomes more disciplined.

Leadership becomes more confident.

. . .

And the organization becomes increasingly prepared to answer the question that defines this entire framework:

If someone asked us today to demonstrate reasonable cybersecurity governance, where would our evidence fall short?

Organizations willing to ask—and honestly answer—that question are already taking one of the most important steps toward becoming truly defensible.

Part V
Governance Under Pressure

Chapter 17
Before an Incident

The best time to strengthen cybersecurity governance is before anyone asks about it.

Unfortunately, many organizations discover weaknesses in governance only after experiencing a significant cyber incident. A ransomware attack exposes gaps in decision-making. A regulatory inquiry reveals incomplete documentation. An insurance claim uncovers inconsistent oversight. Litigation highlights missing evidence that leadership assumed existed.

By then, governance has entered the realm of hindsight.

The opportunity to create evidence has passed.

This chapter is about preventing that situation.

It is about preparing leadership before the Monday Morning Question is ever asked.

Governance Is Built During Quiet Times

There is an old saying in emergency management:

You do not rise to the occasion. You fall to the level of your preparation.

The same principle applies to governance.

Organizations do not suddenly become well governed during a cyber crisis.

They demonstrate the governance they have already established.

The Board cannot invent oversight processes during an emergency.

Executives cannot create decision authority while responding to ransomware.

Risk acceptance cannot be documented after the decision has already been challenged.

Governance is built during ordinary business operations.

. . .

The incident merely reveals its maturity.

Preparation Begins with Leadership

Technology teams spend enormous effort preparing systems.

They patch vulnerabilities.

Deploy security tools.

Monitor networks.

Test backups.

Leadership must prepare with equal discipline.

Preparation includes establishing:

- Clear governance authority.
- Board reporting expectations.
- Executive accountability.
- Risk acceptance processes.
- Escalation criteria.
- Decision rights.
- Documentation standards.
- Independent assurance activities.

These are governance preparations.

They determine how leadership will function when technology alone is no longer enough.

Ask the Monday Morning Question Today

One of the simplest exercises leadership can perform requires no technology.

Gather the Board, executive leadership, cybersecurity leadership, legal counsel, internal audit, and enterprise risk management.

Then ask one question.

> **"If we experienced a significant cyber incident tomorrow morning, what evidence would demonstrate that we exercised reasonable cybersecurity governance?"**

Pause.

Allow the silence.

The first answers usually involve:

. . .

Policies.

Risk assessments.

Board presentations.

Audit reports.

Soon the discussion becomes more thoughtful.

Can we locate them?

Who owns them?

Are they current?

Would they withstand scrutiny?

What evidence is missing?

That conversation alone often identifies governance improvements that years of technical investment never revealed.

Review Governance Before Others Do

Organizations routinely conduct vulnerability assessments against their technology.

Few perform vulnerability assessments against their governance.

Leadership should periodically evaluate:

Does the Board receive meaningful cybersecurity reporting?

Are governance responsibilities clearly documented?

Are significant risks formally accepted?

Do committee charters reflect actual practice?

Are corrective actions consistently tracked?

Does independent assurance occur regularly?

Is governance evidence organized?

Can it be produced quickly?

. . .

These questions should become part of the organization's annual governance review.

Finding weaknesses internally is far preferable to having them discovered externally.

Validate Decision-Making

Preparation is not simply about documentation.

It is about validating governance.

Tabletop exercises provide one of the most effective opportunities to evaluate leadership.

Not merely technical response.

Leadership response.

Can executives prioritize competing objectives?

Do communication procedures work?

Does legal counsel become involved at the appropriate time?

. . .

Are regulators considered?

Does the Board receive the information it needs?

Who authorizes extraordinary expenditures?

Who approves public statements?

Every exercise should evaluate governance as carefully as operational execution.

Prepare the Board

Many Boards receive cybersecurity briefings.

Far fewer receive governance education.

Directors should understand:

Their oversight responsibilities.

How cyber risk differs from operational security.

. . .

What questions they should ask.

What evidence demonstrates reasonable governance.

How governance influences fiduciary responsibility.

Education prepares Directors before decisions become time critical.

Prepared Boards govern better.

Establish the Governance Evidence Repository

Organizations should not wait until after an incident to organize governance evidence.

By then, urgency replaces discipline.

The Governance Evidence Repository should already exist.

Evidence should already be:

- Organized.
- Indexed.
- Protected.
- Searchable.
- Current.

- Accessible.

The repository becomes one of leadership's most valuable assets during an incident because it enables rapid production of governance records when questions inevitably arise.

Test the Repository

Creating a repository is only the first step.

Leadership should periodically test it.

Choose a realistic scenario.

Ask questions such as:

Show every Board discussion regarding ransomware during the past two years.

Produce every risk acceptance decision involving third-party vendors.

Locate all tabletop exercise reports completed within the last eighteen months.

Identify every internal audit finding related to privileged access.

. . .

Measure not only whether the evidence exists.

Measure how long it takes to find it.

Accessibility is an indicator of governance maturity.

Evidence hidden in dozens of disconnected systems provides limited value during a crisis.

Build Governance Into Daily Operations

The strongest governance programs rarely appear extraordinary.

They become ordinary.

Board agendas routinely include cyber oversight.

Risk discussions occur naturally.

Policies are reviewed according to schedule.

Corrective actions remain visible.

. . .

Leadership decisions are documented as they occur.

Independent assurance becomes expected.

Evidence accumulates quietly.

Organizations that normalize governance rarely experience frantic documentation efforts after an incident because governance has already been producing evidence every day.

Avoid the Last-Minute Scramble

Perhaps the greatest governance mistake organizations make is believing they can prepare after the fact.

They schedule emergency policy reviews.

Reconstruct meeting minutes.

Search email archives.

Attempt to recreate undocumented decisions.

Develop governance procedures while regulators are already asking questions.

. . .

These efforts rarely succeed.

They consume valuable time.

They reduce credibility.

They often reveal weaknesses that disciplined governance would have prevented.

Preparation eliminates the scramble.

Confidence Comes From Preparation

Preparation creates confidence.

Not confidence that no cyber incident will occur.

That confidence would be misplaced.

Rather, confidence that leadership will be able to demonstrate disciplined governance when an incident inevitably receives external scrutiny.

Confidence that decisions were documented.

. . .

Confidence that oversight occurred.

Confidence that accountability was clear.

Confidence that governance evidence exists.

Confidence that leadership can answer difficult questions with facts rather than recollections.

The Best Day to Prepare

Organizations frequently ask when they should begin implementing the Defensible Evidence Framework™.

The answer is simple.

Before the next Board meeting.

Before the next risk review.

Before the next tabletop exercise.

Before the next audit.

Before the next cyber incident.

. . .

Governance becomes increasingly difficult to reconstruct as time passes.

Evidence becomes increasingly valuable as time passes.

The best day to prepare was when cybersecurity first became an enterprise risk.

The second-best day is today.

Because governance built before an incident becomes the evidence that protects leadership after one.

Chapter 18
During an Incident

Every cybersecurity incident begins as an operational event.

Very quickly, it becomes a governance event.

The first few hours of a cyber incident are dominated by technical activity. Security teams investigate alerts. Incident responders isolate affected systems. Infrastructure teams work to contain damage. Forensic specialists begin collecting evidence. Communications personnel prepare internal messaging.

These activities are essential.

They are not enough.

At the same time the technical response unfolds, leadership begins making governance decisions that may later receive as much scrutiny as the attack itself.

Who was informed?

. . .

When?

What decisions were made?

Who made them?

What information was available at the time?

What alternatives were considered?

What evidence demonstrates those decisions?

This chapter focuses on the governance of a cyber incident—not the technical response to one.

The Incident Response Plan Is Not the Governance Plan

Most organizations maintain an Incident Response Plan.

Far fewer maintain an Incident Governance Plan.

The distinction matters.

. . .

An Incident Response Plan explains how the organization investigates, contains, eradicates, and recovers from cyber events.

An Incident Governance Plan explains how leadership governs the organization while those technical activities occur.

Governance addresses questions such as:

Who declares a major incident?

When is executive leadership notified?

When is the Board informed?

Who authorizes extraordinary expenditures?

Who approves public statements?

Who determines whether regulatory notification is required?

Who accepts operational risk during recovery?

Who documents leadership decisions?

. . .

Technical response and governance response must occur simultaneously.

Neither replaces the other.

Slow Down to Speed Up

The pressure of a cyber incident often drives organizations toward rapid decision-making.

Speed is important.

Discipline is more important.

One poorly informed executive decision can create greater long-term consequences than the technical incident itself.

Governance during an incident should follow a disciplined rhythm:

Establish facts.

Identify assumptions.

Evaluate available options.

. . .

Document the decision.

Assign accountability.

Review the outcome.

This sequence may require only minutes.

Those few minutes often determine whether leadership appears deliberate or reactive when decisions are later examined.

Facts Versus Assumptions

Every cyber incident begins with uncertainty.

Initial reports are incomplete.

Systems may be unavailable.

Threat intelligence evolves.

Business impacts change rapidly.

Leadership must distinguish between facts and assumptions.

. . .

Good governance requires decision-makers to ask:

What do we know?

What do we believe?

What remains unknown?

What information are we waiting to receive?

What assumptions are driving this decision?

Documenting this distinction becomes valuable evidence later.

It demonstrates that leadership acted responsibly despite incomplete information.

No organization possesses perfect information during a crisis.

Reasonable governance recognizes that reality.

Decision Logs Become Critical Evidence

Few governance artifacts become more valuable than a well-maintained decision log.

. . .

Every significant leadership decision should be recorded contemporaneously.

Each entry should capture:

The decision.

The time.

The participants.

The information available.

The rationale.

Assigned responsibilities.

Required follow-up.

Decision logs often become the most reliable chronology of leadership during an incident.

. . .

Months later, they explain not only what decisions were made, but why they were made.

That context cannot be recreated accurately after memories fade.

The Board Needs Governance Information

Boards do not need operational play-by-play updates.

They need governance information.

Directors should understand:

The nature of the incident.

Business impact.

Known risks.

Legal implications.

Regulatory considerations.

Executive actions.

. . .

Decision points requiring Board oversight.

Expected next steps.

The Board's role is not to direct technical operations.

Its role is to exercise oversight.

Providing governance-focused information enables Directors to fulfill that responsibility without becoming operational managers.

Communications Are Governance Decisions

During a cyber incident, nearly every communication carries governance implications.

Internal communications.

Customer notifications.

Regulatory disclosures.

Insurance reporting.

Public statements.

. . .

Media interviews.

Each communication reflects leadership judgment.

Governance requires consistency.

Leadership should ensure that technical findings, legal advice, executive messaging, and public statements align with one another.

Conflicting communications often become evidence of fragmented governance.

Disciplined communications demonstrate coordinated leadership.

Legal Counsel Is Part of Governance

Legal counsel should not become involved only after difficult questions arise.

Counsel contributes to governance throughout the incident.

Advising leadership on:

Regulatory obligations.

. . .

Preservation of evidence.

Privilege considerations.

Disclosure requirements.

Contractual responsibilities.

Litigation risk.

Legal counsel strengthens governance by helping leadership understand both the legal consequences of decisions and the importance of preserving governance evidence as those decisions are made.

Preserve Governance Evidence While It Is Being Created

Technical evidence receives immediate attention during incident response.

System images.

Log files.

. . .

Forensic artifacts.

Network captures.

Governance evidence deserves equal attention.

Board notifications.

Executive briefings.

Decision logs.

Meeting notes.

Risk discussions.

Approval records.

Communications.

These artifacts should be preserved as part of the incident record.

. . .

Organizations often focus intensely on preserving forensic evidence while overlooking the governance evidence that later explains leadership's actions.

Both are important.

One explains the attack.

The other explains the governance.

Avoid Governance Drift

As incidents continue over several days, governance sometimes begins to deteriorate.

Meetings become informal.

Decisions move into hallway conversations.

Documentation declines.

Approvals become verbal.

Responsibilities blur.

. . .

This phenomenon—governance drift—creates significant evidentiary risk.

Leadership should consciously maintain governance discipline throughout the incident.

Continue documenting.

Continue assigning accountability.

Continue tracking decisions.

Continue following escalation procedures.

The pressure of the incident should strengthen governance—not weaken it.

Recovery Is Also Governance

Organizations often declare victory once systems return to operation.

Governance continues.

Recovery decisions include:

. . .

Returning systems to production.

Accepting temporary risks.

Prioritizing restoration.

Authorizing emergency expenditures.

Communicating with stakeholders.

Monitoring residual risk.

These decisions deserve the same governance discipline as the initial response.

Recovery marks the transition from crisis management back to normal governance—not the end of governance itself.

Remember the Audience

During an incident, leadership naturally focuses on today's operational needs.

Governance requires considering tomorrow's questions as well.

. . .

Those questions may come from:

The Board.

Regulators.

Insurance carriers.

Customers.

Shareholders.

Auditors.

Investigators.

Courts.

Each will eventually ask some variation of the same question:

How did leadership govern the organization during the incident?

. . .

Every governance decision made today becomes part of tomorrow's answer.

The Incident Will End

Every cyber incident eventually concludes.

The systems recover.

Operations stabilize.

The media moves on.

The attackers disappear.

The governance record remains.

Years later, investigators may reconstruct leadership's actions using nothing more than the evidence created during those stressful days.

That evidence will tell one of two stories.

Either governance remained disciplined despite uncertainty...

. . .

...or uncertainty overwhelmed governance.

The difference rarely depends on technical capability.

It depends on leadership.

Because during a cyber incident, technology protects systems.

Governance protects the organization.

And the evidence created during those critical hours becomes the lasting record of how leadership fulfilled that responsibility.

Chapter 19
After an Incident

The incident may be over.

Governance is not.

Many organizations view recovery as the finish line. Systems have been restored. Business operations have resumed. Customers have been notified. Insurance claims have been filed. The crisis appears to be behind them.

From a governance perspective, however, one of the most important phases has only just begun.

The period immediately following a cyber incident determines whether the organization merely recovers—or becomes stronger because of the experience.

The Defensible Evidence Framework™ views every cyber incident as both an operational event and a governance learning opportunity.

. . .

The incident tested the organization.

The aftermath should improve it.

Recovery Is Not the End

Technical recovery and governance recovery are different.

Technical recovery restores systems.

Governance recovery restores confidence.

Leadership should not assume that returning systems to production concludes the organization's responsibilities.

Stakeholders still have questions.

The Board requires assurance.

Regulators may seek additional information.

Insurers will evaluate claims.

Customers may expect greater transparency.

. . .

Internal audit will assess the response.

The organization itself should ask the most important question of all:

What did we learn?

Conduct a Governance Review

Every significant cyber incident should conclude with a structured governance review.

Not a technical review.

A governance review.

Leadership should evaluate:

- Were governance responsibilities clear?
- Did escalation procedures work?
- Was the Board informed appropriately?
- Did executives receive timely information?
- Were decision authorities understood?
- Were communications coordinated?
- Was governance evidence preserved?
- Did oversight function as intended?

The objective is not to assign blame.

It is to evaluate governance performance while memories remain fresh and evidence remains available.

Separate Outcomes from Decisions

One of the greatest dangers following an incident is hindsight bias.

Once the outcome is known, earlier decisions often appear obvious.

They rarely were.

Leadership should evaluate decisions based upon the information available at the time they were made—not information discovered days or weeks later.

A reasonable decision can produce an unfavorable outcome.

An unreasonable decision can occasionally produce a favorable one.

Governance should evaluate the quality of the decision-making process rather than judging every decision solely by its outcome.

This distinction is essential to fair governance.

Update the Evidence

A cyber incident generates a significant amount of governance evidence.

Decision logs.

Board briefings.

Executive communications.

Risk assessments.

After-action reports.

Corrective action plans.

Audit observations.

Lessons learned.

These artifacts should not remain scattered across email systems, collaboration platforms, and individual file repositories.

. . .

They should be incorporated into the Governance Evidence Repository.

Future governance depends upon preserving today's evidence.

The incident becomes part of the organization's institutional memory.

Conduct an After-Action Review

An effective After-Action Review asks four straightforward questions.

What happened?

Establish an objective chronology of the incident.

Separate verified facts from assumptions.

What went well?

Identify governance practices that contributed positively to the response.

Successful governance deserves reinforcement.

. . .

What did not go well?

Examine weaknesses honestly.

Avoid defensiveness.

Governance improves only when leadership acknowledges opportunities for improvement.

What will we change?

Every lesson learned should produce an action.

Every action should have an owner.

Every owner should have a completion date.

Lessons without implementation are merely observations.

Revisit Risk

Every significant incident changes the organization's understanding of risk.

. . .

Previously accepted assumptions may no longer be valid.

Threat likelihood may increase.

Business impacts may be greater than anticipated.

Recovery capabilities may require improvement.

Vendor dependencies may become more significant.

Leadership should revisit the enterprise risk register while these insights remain current.

Risk governance is dynamic.

Every incident provides new information.

Reevaluate Policies

Incidents frequently expose governance assumptions embedded within policies.

Perhaps escalation thresholds proved unrealistic.

. . .

Perhaps reporting timelines were unclear.

Perhaps executive authority required clarification.

Perhaps regulatory notification procedures need refinement.

Policies should evolve based upon operational experience.

Governance that never changes eventually becomes governance that no longer reflects reality.

Verify Corrective Actions

One of the most common governance failures occurs after the After-Action Review.

Corrective actions are identified.

Everyone agrees.

Responsibility is assigned.

Months later, little has changed.

. . .

The incident becomes another report on a shelf.

Leadership must resist this pattern.

Every corrective action should be:

- Assigned.
- Prioritized.
- Funded.
- Tracked.
- Verified.
- Closed.

Independent assurance should confirm completion whenever appropriate.

Improvement—not documentation—is the objective.

Strengthen Board Oversight

Every significant cyber incident presents an opportunity for the Board to improve its governance.

Directors should ask:

What information did we wish we had received earlier?

. . .

Did our reporting support meaningful oversight?

Were governance responsibilities sufficiently clear?

Should committee responsibilities change?

Does Board education need improvement?

Should governance metrics evolve?

Incidents often improve Board effectiveness because they transform theoretical governance discussions into practical leadership experience.

Strengthen Organizational Resilience

Governance maturity is measured by adaptation.

Organizations that merely recover remain vulnerable.

Organizations that learn become resilient.

Resilience means:

. . .

Improved governance.

Improved communication.

Improved decision-making.

Improved preparedness.

Improved evidence.

Every incident should leave the organization stronger than it was before.

Otherwise, the incident becomes an expensive lesson that was never fully learned.

Share the Lessons

Governance improvements should not remain confined to the cybersecurity team.

Executive leadership should understand them.

The Board should review them.

. . .

Business leaders should incorporate them into planning.

Internal audit should consider them in future assessments.

Enterprise Risk Management should update enterprise risk profiles.

Lessons become organizational assets only when they are shared.

The Incident Becomes Governance History

Over time, every significant cyber incident becomes part of the organization's governance history.

Future Boards will review it.

New executives will learn from it.

Auditors will reference it.

Regulators may examine it.

The incident therefore becomes more than an operational event.

It becomes evidence of how the organization governs adversity.

. . .

That history should demonstrate thoughtful leadership rather than organizational confusion.

The Strongest Organizations Improve

The strongest organizations are not those that never experience cyber incidents.

Such organizations do not exist.

The strongest organizations are those that consistently emerge from incidents with stronger governance than they had before.

They update policies.

Improve oversight.

Clarify accountability.

Strengthen assurance.

Enhance evidence.

Refine decision-making.

. . .

They treat every incident as an investment in future governance maturity.

Preparing for the Next Monday Morning

Every cyber incident ends.

Another will eventually begin.

The question is whether the organization will govern the next one better than the last.

The answer depends entirely on what leadership does after the current incident concludes.

Organizations that preserve evidence, evaluate governance honestly, implement meaningful improvements, and strengthen leadership emerge better prepared for the future.

Those that simply return to business as usual remain vulnerable to repeating the same governance mistakes.

The true measure of a cyber incident is therefore not how quickly systems were restored.

It is whether the organization's governance became stronger because the incident occurred.

. . .

Because in the end, every incident writes another chapter in the organization's governance story.

Leadership decides whether that chapter becomes a lesson...or a warning.

Part VI
Defensible Governance in Practice

Chapter 20
The Apex Case Study

Theory explains governance.

Experience proves it.

Throughout this book, we have examined the principles of the Defensible Evidence Framework™ through individual examples. This chapter brings those principles together in a fictional—but entirely realistic—case study illustrating how governance evidence influences the outcome of a significant cyber incident.

The organizations described here are fictional.

The governance challenges are not.

Every decision, document, and governance activity presented in this chapter reflects situations organizations encounter every day.

The purpose of this case study is not to demonstrate perfect cybersecurity.

. . .

It is to demonstrate the difference between an organization that governs cybersecurity through evidence and one that governs primarily through assumption.

Meet Apex Manufacturing

Apex Manufacturing is a mid-sized industrial manufacturer with approximately 2,800 employees operating across four states. Its products support the aerospace, transportation, and energy sectors.

Like many successful manufacturers, Apex has invested heavily in operational technology (OT), cloud services, enterprise resource planning systems, and connected manufacturing equipment.

Its Board receives quarterly cybersecurity reports.

Executive leadership recognizes cybersecurity as an enterprise risk.

The company maintains cyber insurance.

Annual penetration testing is performed.

Internal Audit periodically reviews cybersecurity governance.

. . .

On paper, Apex appears well prepared.

Three Years Earlier

Three years before the incident described in this chapter, Apex's Board approved a revised Cybersecurity Governance Charter.

The charter established:

- Board oversight responsibilities.
- Executive accountability.
- Quarterly cyber risk reporting.
- Annual governance reviews.
- Independent assurance requirements.
- Formal cyber risk acceptance procedures.

These documents became the foundation of Apex's Governance Evidence.

Leadership did not know whether the organization would ever experience a significant cyber incident.

They governed as though one eventually would.

Year Two

During the following year, enterprise risk assessments identified increasing dependence upon third-party software suppliers.

. . .

Rather than treating the issue as an information technology concern, executive leadership elevated supplier cyber risk to the enterprise risk register.

The Board requested additional reporting.

Procurement strengthened vendor assessments.

Internal Audit added third-party governance to its audit plan.

Evidence accumulated naturally.

Board minutes.

Risk registers.

Committee reports.

Vendor assessments.

Audit plans.

No one created these records because litigation was anticipated.

. . .

They simply reflected governance occurring as intended.

Year One

The following year, Apex conducted a ransomware tabletop exercise.

Participants included:

The Chief Executive Officer.

The Chief Financial Officer.

The Chief Information Security Officer.

General Counsel.

Operations leadership.

Communications.

Human Resources.

The Board's Risk Committee Chair.

. . .

The exercise revealed several governance weaknesses.

Escalation procedures required clarification.

Board notification thresholds were inconsistent.

Executive decision authority needed refinement.

Insurance notification procedures required updating.

Rather than filing the exercise report away, leadership assigned corrective actions.

Each action received:

An owner.

A completion date.

Executive oversight.

Verification.

. . .

Six months later, Internal Audit confirmed every corrective action had been completed.

Additional evidence was created.

Monday Morning

At 3:12 a.m. on a Monday morning, Apex detected unusual activity affecting several manufacturing systems.

Within thirty minutes, the Incident Response Team declared a major cyber incident.

By 5:00 a.m., executive leadership had assembled.

At 6:15 a.m., the CEO briefed the Board Chair.

By 8:00 a.m., the full Board received its initial governance briefing.

The technical response began immediately.

So did governance.

Leadership During the Incident

Over the next seventy-two hours, Apex leadership made dozens of governance decisions.

. . .

Production was temporarily suspended at two facilities.

Outside forensic specialists were retained.

Cyber insurance was notified.

Law enforcement was contacted.

Customers received coordinated communications.

The Board met twice.

Every significant decision entered the incident decision log.

Every executive briefing was preserved.

Every Board update became part of the governance record.

Leadership understood that today's decisions would become tomorrow's evidence.

Regulatory Review

Several weeks later, regulators requested documentation demonstrating Apex's cybersecurity governance.

The request included:

Board oversight.

Risk assessments.

Governance policies.

Incident decision records.

Corrective action history.

Executive reporting.

Independent assurance.

Unlike many organizations, Apex did not begin searching through email archives.

. . .

Its Governance Evidence Repository already contained the required records.

Within hours, leadership produced:

Board minutes.

Committee charters.

Governance policies.

Risk registers.

Tabletop exercise reports.

Audit findings.

Corrective action verification.

Decision logs.

Executive briefings.

. . .

Evidence preservation records.

The documentation demonstrated that governance had existed long before the incident occurred.

The Insurance Review

The cyber insurance carrier conducted its own review.

Investigators examined not only the technical incident but also Apex's governance.

Questions included:

Did leadership understand cyber risk?

Was the Board engaged?

Were known risks addressed?

Did governance improve over time?

The evidence answered each question.

. . .

The insurer noted that although the attack had been successful, Apex demonstrated disciplined governance before, during, and after the incident.

That distinction influenced both the claim process and subsequent underwriting discussions.

The Board's Perspective

Several months after recovery, the Board conducted its annual governance review.

The discussion focused less on the attacker than on leadership.

Which governance decisions worked well?

Where did reporting require improvement?

Were escalation thresholds appropriate?

How could oversight become even stronger?

The Board concluded that governance had functioned effectively—but not perfectly.

. . .

Additional improvements were approved.

Governance continued to evolve.

A Different Organization

Now imagine the same attack occurring at another manufacturer.

The technology is similar.

The ransomware behaves the same way.

The operational disruption is nearly identical.

The governance is not.

The Board rarely discusses cybersecurity.

Risk acceptance occurs informally.

Committee responsibilities are unclear.

Policies are outdated.

. . .

Incident response has never been exercised.

Audit findings remain unresolved.

Board minutes contain little evidence of discussion.

Decision logs do not exist.

Following the incident, leadership attempts to reconstruct governance.

Meeting notes are incomplete.

Memories conflict.

Documentation cannot be located.

Investigators receive explanations rather than evidence.

The technical attack may be identical.

The governance outcome is entirely different.

The Difference Was Never the Technology

Both organizations experienced a serious cyber incident.

Both restored operations.

Both incurred financial loss.

Both suffered business disruption.

Yet one organization demonstrated reasonable governance.

The other struggled to demonstrate governance at all.

The difference was not the firewall.

Not endpoint protection.

Not vulnerability management.

Not artificial intelligence.

The difference was governance.

. . .

More specifically, the difference was evidence.

Lessons from Apex

The Apex case study illustrates several principles that appear throughout this book.

Governance begins long before an incident.

Evidence is created naturally through disciplined leadership.

Risk governance requires executive engagement.

Board oversight must be continuous.

Corrective actions matter only when verified.

Repositories are built before they are needed.

Decision logs become invaluable.

Assurance validates confidence.

Most importantly, governance should never depend upon memory.

. . .

It should depend upon evidence.

The Story Every Organization Wants to Tell

Every significant cyber incident eventually becomes a story.

Sometimes the organization tells it.

Sometimes regulators tell it.

Sometimes insurers tell it.

Sometimes courts tell it.

The strongest organizations ensure that story is supported by evidence created long before anyone asks for it.

That is the lesson of Apex Manufacturing.

Not that good governance prevents every cyber incident.

It does not.

Rather, good governance ensures that when the incident is over, leadership can demonstrate that it acted with clarity, accountability, and reasonable care.

That is the essence of the Defensible Evidence Framework™.

And that is what makes governance truly defensible.

Chapter 21
The Organization That Couldn't Answer

Every governance failure begins with a question.

Not a sophisticated technical question.

Not a complex forensic question.

Usually, it is something remarkably simple.

"Can you show us?"

Can you show us where the Board discussed this risk?

Can you show us who accepted the residual risk?

Can you show us the policy that governed this decision?

. . .

Can you show us the corrective action?

Can you show us the evidence?

For many organizations, those questions become more difficult than recovering encrypted servers.

Because while systems can often be restored from backups, governance cannot be restored from memory.

The Call No Leader Wants

At 9:30 on a Tuesday morning, the General Counsel received a call.

The organization's cyber insurer requested documentation supporting the company's cybersecurity governance during the two years preceding a recent ransomware incident.

The request appeared straightforward.

Board oversight documentation.

Risk assessments.

Governance policies.

. . .

Incident preparedness.

Corrective action history.

Executive reporting.

Leadership assumed gathering the documents would take a few hours.

It took six weeks.

"We Know We Did It"

The search began with confidence.

"We know the Board discussed cybersecurity."

"We definitely reviewed that policy."

"I'm certain we approved that funding."

"We had several meetings about vendor security."

. . .

Everyone remembered governance.

No one could consistently prove it.

Board minutes mentioned cybersecurity only briefly.

The detailed presentations had never been preserved.

Risk registers existed in multiple versions.

No one knew which version represented the official record.

Policies had been updated repeatedly.

Approval records were incomplete.

Meeting notes remained in personal notebooks.

Corrective actions existed in spreadsheets maintained by different departments.

The governance had occurred.

. . .

The evidence had not survived.

A Patchwork of Information

As more documents surfaced, another problem emerged.

Nothing connected.

Risk assessments did not reference Board discussions.

Board minutes did not reference executive reports.

Executive reports did not reference corrective actions.

Audit findings existed without documented closure.

Policies referenced committees that no longer existed.

Leadership could produce documents.

It could not produce a governance story.

Every artifact appeared isolated.

. . .

Investigators had to assemble the narrative themselves.

Organizations rarely benefit when others reconstruct their governance.

The Missing Decisions

One issue attracted particular attention.

Two years earlier, management had decided to delay implementation of multifactor authentication for a critical legacy environment because of operational concerns.

The decision itself was reasonable.

The documentation was not.

No record identified:

Who made the decision.

Who participated.

What alternatives were considered.

. . .

What residual risk remained.

When the decision would be revisited.

Months later, investigators knew the outcome.

They did not know the reasoning.

Without evidence, the decision appeared arbitrary.

The Board Asked the Wrong Question

During a special Board meeting following the incident, one Director asked:

> "Why weren't we told this could happen?"

The Chief Information Security Officer responded carefully.

> "You were."

Silence filled the room.

The Board had indeed received quarterly cybersecurity reports.

Unfortunately, the reports emphasized technical activity rather than governance decisions.

Threats.

Patch metrics.

Training completion.

Security alerts.

Very little discussion connected those operational activities to enterprise risk, governance priorities, or Board decision-making.

Leadership had communicated cybersecurity.

It had not communicated governance.

Internal Audit Looked Back

Several weeks after the incident, Internal Audit began reviewing historical governance.

Its findings were revealing.

Cybersecurity policies had been reviewed inconsistently.

Committee responsibilities overlapped.

Risk acceptance lacked executive approval.

Corrective actions remained open for years.

Board education occurred sporadically.

Governance evidence was dispersed across more than a dozen systems.

None of these findings caused the ransomware attack.

Collectively, however, they made it extraordinarily difficult to demonstrate reasonable governance.

. . .

The organization had not merely experienced a cyber incident.

It had experienced an evidence failure.

The Investigation Changed

As the investigation continued, technical questions gradually diminished.

Governance questions increased.

Investigators wanted to understand leadership.

How were cyber risks communicated?

Who evaluated them?

How were priorities established?

How did the Board exercise oversight?

How did management verify implementation?

How did governance improve over time?

. . .

The organization struggled because it had prepared to defend technology.

It had never prepared to defend governance.

The Cost of Uncertainty

No investigator accused leadership of acting recklessly.

No evidence suggested intentional misconduct.

The problem was uncertainty.

Missing documentation created unanswered questions.

Unanswered questions required additional investigation.

Additional investigation increased legal costs.

Insurance reviews took longer.

Regulatory discussions expanded.

Executive time shifted from leading recovery to searching for historical records.

. . .

The absence of evidence did not prove governance failed.

It prevented leadership from proving governance succeeded.

The Turning Point

Months after the incident, the Board commissioned an independent governance review.

The consultants delivered an observation that surprised everyone.

> "You don't have a cybersecurity problem."
>
> "You have a governance evidence problem."

The organization possessed capable security professionals.

Competent leadership.

Reasonable technical controls.

Dedicated employees.

. . .

What it lacked was a disciplined system for preserving the evidence created through governance.

The consultants recommended implementing what would later become known within the organization as an evidence-first governance model.

Building a Different Future

Over the next eighteen months, the organization fundamentally changed how it governed cybersecurity.

Governance responsibilities were clarified.

Board reporting shifted from technical metrics to governance decisions.

Risk acceptance became formal.

Committee charters were revised.

Decision logs became standard practice.

The Governance Evidence Repository was established.

. . .

Internal Audit expanded governance assurance.

Evidence Gap Analyses became annual exercises.

None of these changes prevented cyber threats.

All of them strengthened governance.

The Next Investigation

Three years later, the organization experienced another significant cybersecurity incident.

The attackers were different.

The technology had changed.

Leadership had changed.

The governance response was remarkably different.

When regulators requested governance documentation, the organization responded within forty-eight hours.

Board oversight records.

. . .

Risk acceptance decisions.

Governance policies.

Tabletop exercises.

Audit reports.

Corrective action verification.

Decision logs.

Executive communications.

The documentation told a complete story.

Not because the organization had become perfectly secure.

Because it had become demonstrably governed.

The Lesson

Every organization believes it will be able to explain its governance.

Many cannot.

The difference rarely becomes apparent until someone asks to see the evidence.

That moment defines the organization.

Some respond with recollections.

Others respond with records.

Some explain what they intended to do.

Others demonstrate what they actually did.

That distinction is the difference between governance that exists...

...and governance that can be defended.

The Question Every Leader Should Ask

The story in this chapter is fictional.

The questions are not.

Every organization will eventually face some version of them.

Whether from a regulator.

An insurer.

An auditor.

A customer.

A Board member.

A shareholder.

Or a court.

The only uncertainty is when.

· · ·

The organizations best prepared for that moment are not those that experienced the fewest cyber incidents.

They are those that ensured every significant governance activity produced credible, organized, and accessible evidence long before anyone asked for it.

Because the most dangerous governance question is rarely the one leadership cannot answer.

It is the one leadership thought it would never be asked.

Part VII
Applying the Framework

Chapter 22
Public Companies

Public companies operate under a governance environment unlike any other. While every organization is expected to exercise reasonable cybersecurity oversight, publicly traded companies must also satisfy securities laws, disclosure obligations, investor expectations, and increasingly active regulatory scrutiny. Cybersecurity governance is no longer evaluated solely by whether an attack occurred, but by whether the Board exercised informed oversight before, during, and after the event.

For public companies, cybersecurity is therefore both a governance responsibility and a market responsibility.

The consequences of governance failure extend well beyond operational disruption. A significant cyber incident can trigger SEC reporting obligations, shareholder litigation, class-action lawsuits, regulatory investigations, credit rating impacts, insurance disputes, executive turnover, and long-term reputational damage. Even organizations that successfully recover operationally may spend years recovering investor confidence.

The Board's responsibility is not to prevent every cyberattack. Its responsibility is to demonstrate that cybersecurity risk has been governed with the same rigor applied to financial reporting, enterprise risk management, and corporate compliance.

Governance Priorities for Public Companies

Effective public company governance should establish:

- Board-level oversight of cybersecurity as an enterprise risk.
- Clearly defined executive accountability for cybersecurity governance.
- Regular reporting using business-oriented metrics rather than technical statistics.
- Integration of cybersecurity into enterprise risk management (ERM).
- Independent assurance of cybersecurity controls and governance processes.
- Formal documentation of Board deliberations and oversight activities.

Cybersecurity should appear as a standing governance agenda item rather than an annual presentation following audit season.

SEC Expectations

Recent SEC cybersecurity disclosure rules significantly changed governance expectations.

Public companies should be prepared to demonstrate:

- how the organization identifies and manages material cyber risk;
- the Board's role in cybersecurity oversight;
- management's responsibilities for assessing and managing cyber risk;
- how cybersecurity risks affect business strategy, operations, and financial condition; and
- processes used to determine whether cyber incidents are material.

These disclosures shift cybersecurity governance from an internal management concern to a matter of public accountability.

Materiality Is a Governance Decision

One of the most difficult responsibilities facing public company leadership is determining whether a cyber incident is material.

Materiality cannot be evaluated solely through technical impact.

Boards should consider:

- financial consequences;
- operational disruption;
- legal exposure;
- regulatory obligations;
- reputational harm;
- effects on customers and shareholders; and
- strategic implications.

. . .

Determining materiality requires coordinated input from executive leadership, legal counsel, finance, investor relations, and cybersecurity leadership. Governance exists to ensure those perspectives are integrated into a defensible decision-making process.

Documentation Becomes Evidence

Public companies should assume that every significant cyber incident may eventually be reviewed by regulators, external auditors, shareholders, or courts.

Accordingly, governance documentation should demonstrate:

- the information presented to leadership;
- questions asked by the Board;
- decisions made;
- rationale supporting those decisions;
- follow-up actions;
- ongoing monitoring; and
- lessons incorporated into governance improvements.

Well-maintained governance records often become the strongest evidence that Directors fulfilled their fiduciary duties.

Cybersecurity and Fiduciary Duty

For public company Directors, cybersecurity oversight directly supports the fiduciary duties of care and loyalty.

. . .

Reasonable oversight includes:

- remaining informed regarding cyber risk;
- ensuring qualified leadership is in place;
- approving governance policies;
- reviewing independent assurance activities;
- challenging management when necessary; and
- documenting oversight activities.

Directors are not expected to be cybersecurity experts. They are expected to govern cybersecurity with the same diligence they apply to every other material enterprise risk.

Applying the Defensible Evidence Framework™

The Defensible Evidence Framework™ provides public companies with a practical structure for demonstrating governance readiness.

Evidence should naturally exist across every governance layer:

Governance

- Board committee charters
- Cybersecurity oversight agendas
- Board minutes
- Governance policies

Risk

- Enterprise cyber risk assessments
- Risk appetite statements
- Risk acceptance decisions
- Executive reporting

Assurance

- Internal audit reports
- Independent assessments
- Penetration testing summaries
- Remediation tracking

Incident Readiness

- Incident response governance plans
- Tabletop exercise reports
- Crisis communication procedures
- Disclosure decision documentation

Continuous Oversight

- Quarterly governance dashboards
- Annual Board education
- Policy reviews
- Lessons-learned reports

Collectively, these artifacts demonstrate that cybersecurity governance is not an occasional activity but an ongoing Board discipline.

Public Company Governance Checklist

Boards should periodically confirm that they can answer "yes" to the following questions:

- Does the Board receive regular cybersecurity governance reporting?
- Is cyber risk integrated into enterprise risk management?
- Are materiality decisions governed through a documented process?
- Are independent assessments performed on a regular basis?
- Are cybersecurity responsibilities clearly assigned?
- Does the Board periodically participate in incident tabletop exercises?
- Are governance decisions documented sufficiently to withstand regulatory scrutiny?
- Is cybersecurity oversight reviewed as part of the Board's annual governance assessment?

Governance Reality

Public markets do not expect organizations to be breach-proof.

They increasingly expect Boards to be governance-ready.

When investors, regulators, or courts evaluate cybersecurity after a significant incident, the defining question is rarely *"Why were you attacked?"* More often it is *"Can you demonstrate that leadership exercised informed, reasonable, and documented oversight?"*

. . .

That is the standard public company Boards must now be prepared to meet.

Chapter 23
Healthcare Organizations

Healthcare organizations occupy one of the most challenging cybersecurity governance environments of any sector. Every cyber incident carries consequences that extend beyond financial loss or regulatory scrutiny. Patient safety, continuity of care, clinical operations, and public trust are all directly affected by the organization's ability to govern cyber risk effectively.

Healthcare cybersecurity is therefore not simply an information technology issue. It is an extension of patient care governance.

Boards governing hospitals, physician practices, health systems, clinics, laboratories, insurers, and other healthcare entities must recognize that cybersecurity failures can quickly become failures of clinical operations. Electronic health records, diagnostic systems, medical devices, pharmacy systems, laboratory systems, imaging platforms, scheduling applications, and communications infrastructure all support the organization's mission of delivering care.

Protecting those systems is inseparable from protecting patients.

Governance Priorities for Healthcare Organizations

Healthcare Boards should ensure cybersecurity governance addresses:

- patient safety;
- continuity of clinical operations;
- protection of electronic protected health information (ePHI);
- regulatory compliance;
- third-party clinical vendors;
- medical device security;
- disaster recovery and business continuity;
- executive decision-making during cyber incidents.

Cybersecurity should be integrated into enterprise governance alongside patient safety, quality improvement, and organizational resilience.

HIPAA Is the Floor—Not the Ceiling

Many healthcare organizations mistakenly equate HIPAA compliance with cybersecurity readiness.

HIPAA establishes minimum administrative, physical, and technical safeguards for protecting protected health information. Compliance with those requirements does not necessarily demonstrate that an organization is prepared to withstand modern ransomware, supply chain attacks, insider threats, or nation-state adversaries.

. . .

Boards should avoid asking:

"Are we HIPAA compliant?"

Instead, they should ask:

"Are we governing cyber risk appropriately?"

Compliance satisfies regulatory obligations.

Governance protects the mission.

Patient Safety Is a Cybersecurity Outcome

Healthcare has become increasingly dependent upon interconnected digital systems.

A successful cyberattack may affect:

- emergency departments;
- operating rooms;
- medication administration;
- laboratory services;
- diagnostic imaging;

- pharmacy operations;
- intensive care units;
- ambulance diversion;
- clinical communications.

When those systems become unavailable, patient outcomes may be directly affected.

For healthcare Boards, cybersecurity oversight is therefore inseparable from oversight of patient safety and quality of care.

Ransomware as a Clinical Crisis

Healthcare has become one of the primary targets for ransomware operators because organizations often face extraordinary pressure to restore clinical operations quickly.

Governance should recognize that ransomware is not merely an IT outage.

It may require leadership decisions involving:

- diversion of patients;
- cancellation of elective procedures;
- manual documentation processes;
- coordination with public health authorities;
- regulatory reporting;
- crisis communications;
- legal counsel;

- law enforcement.

Boards should ensure these governance decisions are exercised before an actual incident through executive tabletop exercises.

Medical Devices Expand the Attack Surface

Modern healthcare depends upon thousands of connected medical devices.

These may include:

- infusion pumps;
- patient monitoring systems;
- imaging equipment;
- laboratory analyzers;
- implant programming devices;
- smart medication cabinets;
- wireless telemetry systems.

Many of these devices have extended service lives, limited patching capabilities, and dependencies on third-party vendors.

Boards should require governance processes that evaluate medical device cybersecurity as part of enterprise risk management rather than treating it solely as a biomedical engineering concern.

Third-Party Clinical Dependencies

Healthcare organizations rely extensively on external service providers.

Critical dependencies often include:

- cloud-hosted electronic health record platforms;
- revenue cycle management vendors;
- medical device manufacturers;
- diagnostic laboratories;
- managed service providers;
- telehealth platforms;
- pharmacy benefit managers;
- outsourced billing services.

A compromise affecting one of these providers can rapidly affect patient care across the organization.

Healthcare governance should therefore include rigorous third-party risk oversight, contractual security requirements, ongoing monitoring, and contingency planning.

Applying the Defensible Evidence Framework™

The Defensible Evidence Framework™ provides healthcare organizations with a structured approach to demonstrating reasonable governance before, during, and after cyber incidents.

. . .

Evidence should naturally exist throughout the governance lifecycle.

Governance

- Board cybersecurity agendas
- Governance committee minutes
- Executive accountability assignments
- Cybersecurity governance policies

Risk

- Enterprise cyber risk assessments
- Clinical systems criticality analysis
- Medical device risk inventories
- Risk acceptance documentation

Assurance

- HIPAA Security Rule assessments
- Internal audit reports
- Independent security assessments
- Penetration testing summaries
- Remediation tracking

Incident Readiness

- Incident response governance plans
- Downtime procedures
- Clinical continuity plans
- Tabletop exercise reports
- Crisis communication plans

Continuous Oversight

- Quarterly Board dashboards
- Policy review cycles
- Annual governance education
- Lessons-learned reviews
- Continuous monitoring reports

Collectively, these governance artifacts demonstrate that cybersecurity oversight is embedded within organizational leadership rather than isolated within the information technology department.

Healthcare Governance Checklist

Boards should periodically ask:

- Is cybersecurity treated as a patient safety issue?
- Does executive leadership regularly brief the Board on cyber risk?
- Have mission-critical clinical systems been identified?
- Are ransomware response procedures regularly exercised?
- Are medical devices included in enterprise cyber risk assessments?
- Is third-party healthcare risk continuously monitored?
- Are HIPAA compliance activities integrated into broader cybersecurity governance?
- Are governance decisions documented to demonstrate reasonable oversight?

Governance Reality

Patients never see the organization's cybersecurity program.

They experience its effectiveness through uninterrupted care, protected health information, reliable clinical systems, and confidence that their healthcare providers can continue delivering services during adverse events.

Healthcare organizations cannot eliminate cyber risk.

They can, however, govern it in ways that protect patients, support clinicians, satisfy regulators, and demonstrate that leadership exercised informed, reasonable, and documented oversight before a crisis occurs.

In healthcare, cybersecurity governance ultimately protects more than information.

It protects lives.

Chapter 24
Financial Services

Financial institutions operate in one of the most heavily regulated and closely scrutinized cybersecurity environments in the world. Banks, credit unions, investment firms, insurance companies, payment processors, fintech organizations, and capital markets firms are entrusted with safeguarding not only sensitive financial information, but also the stability of the financial system itself.

Cybersecurity governance within financial services extends beyond protecting assets.

It preserves trust.

Customers trust that their money is available when needed. Markets trust that transactions are accurate. Regulators trust that institutions are operating safely and soundly. Investors trust that leadership is exercising prudent oversight over technology and cyber risk.

That trust is earned through governance.

. . .

And governance is demonstrated through evidence.

Governance Is a Safety-and-Soundness Issue

Financial institutions have long understood that operational resilience is fundamental to their mission.

Cybersecurity is now inseparable from operational resilience.

A cyber incident can interrupt payment systems, online banking, securities trading, wire transfers, loan processing, fraud detection, customer authentication, and regulatory reporting.

As a result, Boards must govern cybersecurity with the same discipline applied to financial controls, liquidity, capital management, and enterprise risk.

Cyber risk is no longer an information technology issue.

It is an enterprise governance issue.

Regulatory Expectations Continue to Increase

Financial institutions operate under numerous regulatory expectations that vary by jurisdiction and organizational type.

. . .

Although specific requirements differ, regulators consistently expect organizations to demonstrate:

- active Board oversight;
- enterprise cyber risk management;
- effective internal controls;
- operational resilience;
- third-party risk management;
- incident response capabilities;
- independent assurance; and
- continual governance improvement.

Increasingly, regulators are interested not only in whether these activities occur, but whether leadership can demonstrate that they occurred through credible evidence.

Governance Begins with the Board

The Board establishes the institution's cyber risk appetite.

It approves governance policies.

It oversees executive leadership.

It reviews enterprise cyber risk.

It evaluates independent assurance.

. . .

It ensures that cybersecurity remains aligned with the institution's strategic objectives.

Directors are not expected to manage security operations.

They are expected to ensure those operations are governed appropriately.

Board discussions should increasingly focus on business resilience, operational risk, customer impact, regulatory obligations, and governance effectiveness rather than technical implementation details.

Cyber Risk Is Enterprise Risk

Financial institutions have historically excelled at managing financial risk.

Cyber risk deserves the same discipline.

Effective governance integrates cybersecurity into enterprise risk management alongside:

- credit risk;
- market risk;
- liquidity risk;
- operational risk;
- legal risk;
- compliance risk;

- reputational risk; and
- strategic risk.

This integration enables leadership to prioritize cybersecurity decisions within the broader context of organizational objectives.

Operational Resilience

Operational resilience has become a defining governance priority for financial institutions.

Boards should ensure the organization can continue delivering critical services despite cyber disruptions.

Governance should address:

- payment processing;
- customer account access;
- treasury operations;
- securities settlement;
- fraud monitoring;
- customer communications;
- business continuity; and
- disaster recovery.

The objective is not merely restoring technology.

. . .

It is maintaining confidence in the institution's ability to operate.

Third-Party and Supply Chain Governance

Modern financial institutions depend upon a complex ecosystem of external providers.

These commonly include:

- cloud service providers;
- core banking platforms;
- payment processors;
- investment platforms;
- fraud detection services;
- managed security providers;
- customer identity platforms; and
- fintech partners.

A disruption affecting one provider can rapidly affect multiple business functions.

Boards should therefore require rigorous governance over third-party risk throughout the vendor lifecycle.

Due diligence should not end when the contract is signed.

Continuous oversight should become part of normal governance.

Incident Governance

During a significant cyber incident, executive leadership must balance technical response with governance responsibilities.

Financial institutions frequently face decisions involving:

- regulatory notifications;
- customer communications;
- law enforcement coordination;
- payment system continuity;
- fraud prevention;
- liquidity impacts;
- reputational risk; and
- market confidence.

These decisions require disciplined governance.

Decision logs, executive briefings, Board communications, and governance timelines become critical evidence demonstrating reasonable leadership under pressure.

Applying the Defensible Evidence Framework™

The Defensible Evidence Framework™ enables financial institutions to organize governance evidence in a manner that supports regulatory examinations, internal oversight, independent assurance, and organizational resilience.

. . .

The Defensible Evidence Framework™

Evidence should naturally exist across each governance domain.

Governance

- Board cybersecurity policies
- Committee charters
- Governance calendars
- Executive accountability assignments
- Cyber risk appetite statements

Risk

- Enterprise cyber risk register
- Operational resilience assessments
- Third-party risk assessments
- Risk acceptance decisions
- Business impact analyses

Oversight

- Board minutes
- Executive dashboards
- Risk committee reports
- Corrective action tracking
- Governance metrics

Operational

- Incident response plans
- Business continuity testing
- Disaster recovery exercises
- Crisis communication procedures

- Tabletop exercise reports

Assurance

- Internal audit reports
- Independent cybersecurity assessments
- Penetration testing
- Regulatory examination responses
- Governance maturity assessments

Together, these artifacts demonstrate that cybersecurity governance is embedded within the institution's overall governance program rather than operating independently.

Financial Services Governance Checklist

Boards should periodically confirm that they can answer "yes" to the following questions:

- Is cybersecurity integrated into enterprise risk management?
- Does the Board regularly review cyber risk and operational resilience?
- Have critical financial services been identified and prioritized?
- Are third-party technology risks continuously monitored?
- Are ransomware and operational disruption scenarios exercised through executive tabletop exercises?
- Are significant governance decisions documented contemporaneously?

- Does Internal Audit evaluate cybersecurity governance on a regular basis?
- Can governance evidence be produced quickly during a regulatory examination or significant incident?

Governance Reality

Customers entrust financial institutions with more than money.

They entrust them with confidence.

That confidence depends upon uninterrupted operations, secure transactions, effective fraud prevention, and responsible leadership.

Technology plays an essential role in protecting financial services.

Governance determines how leadership prepares for disruption, responds under pressure, and demonstrates accountability afterward.

Financial institutions that embrace evidence-driven governance position themselves not only to satisfy regulators, but also to strengthen resilience, reinforce stakeholder confidence, and preserve trust when it matters most.

In financial services, trust is the institution's most valuable asset.

. . .

Cybersecurity governance is one of the primary ways leadership protects it.

Chapter 25
Government

Government organizations occupy a unique position in cybersecurity governance. Unlike private-sector organizations that primarily answer to shareholders and customers, government agencies answer to citizens, elected officials, oversight bodies, inspectors general, auditors, funding authorities, and, ultimately, the public.

Their mission is not measured by profit.

It is measured by public trust.

Every cyber incident affecting a government organization has the potential to disrupt essential public services, expose sensitive information, delay critical operations, erode confidence in public institutions, and compromise national, state, or local interests.

For government leaders, cybersecurity governance is therefore a matter of public stewardship.

The question is not simply whether systems remained secure.

. . .

The question is whether public officials exercised reasonable care in protecting the services entrusted to them.

Governance Protects the Public Mission

Government organizations perform missions that citizens depend upon every day.

These may include:

- public safety;
- healthcare;
- education;
- transportation;
- emergency management;
- taxation;
- elections;
- social services;
- justice systems;
- national defense.

Cybersecurity failures can interrupt these services at precisely the moment they are needed most.

Boards, governing bodies, agency heads, commissioners, elected officials, and executive leadership therefore share responsibility for governing cyber risk as part of their broader duty to serve the public.

Accountability Extends Beyond Compliance

Government organizations often operate within extensive statutory and regulatory requirements.

Compliance is essential.

Governance requires more.

Public officials must demonstrate:

- responsible stewardship of public resources;
- informed decision-making;
- effective oversight;
- transparent accountability;
- continuous improvement; and
- appropriate management of public risk.

Compliance answers whether requirements were followed.

Governance answers whether leadership fulfilled its public responsibilities.

Governance Must Survive Leadership Changes

Unlike many private organizations, government agencies experience frequent leadership transitions.

. . .

New administrations arrive.

Agency heads change.

Department directors retire.

Elected officials leave office.

Cyber threats do not pause during these transitions.

Governance must therefore be institutional rather than personal.

Policies, oversight structures, governance records, and evidence repositories preserve continuity regardless of changes in political or executive leadership.

Institutional memory becomes one of government's greatest governance assets.

Public Trust Is a Governance Outcome

Citizens rarely evaluate government cybersecurity by examining technical controls.

. . .

They evaluate whether government continues delivering services during disruption.

Public trust depends upon:

- reliable services;
- transparent communication;
- responsible leadership;
- accountable decision-making; and
- effective recovery.

Good governance strengthens confidence long after a cyber incident has concluded.

Poor governance can undermine confidence for years.

Public Funding Requires Public Accountability

Government organizations are entrusted with taxpayer resources.

Cybersecurity investments therefore require disciplined governance.

Leadership should be prepared to demonstrate:

- why investments were made;
- how priorities were established;
- how risks were evaluated;

- how funding decisions supported organizational objectives;
- how effectiveness was measured; and
- how public resources were managed responsibly.

Evidence documenting these decisions becomes increasingly important during legislative inquiries, budget reviews, audits, and public records requests.

Grants and Federal Funding

Many government agencies receive state or federal funding that carries cybersecurity governance expectations.

Grant recipients may be required to demonstrate:

- risk management activities;
- internal controls;
- oversight mechanisms;
- procurement governance;
- documentation of expenditures;
- independent audits; and
- continuous monitoring.

The Defensible Evidence Framework™ complements these expectations by organizing governance evidence into a coherent body of proof.

. . .

Rather than producing documentation only when grant reviews occur, agencies continuously create evidence as governance takes place.

Incident Governance in the Public Sector

Government cyber incidents often involve more stakeholders than comparable private-sector events.

Leadership may need to coordinate with:

- elected officials;
- law enforcement;
- emergency management agencies;
- state cybersecurity offices;
- federal partners;
- oversight committees;
- inspectors general;
- public information officers;
- external auditors; and
- the media.

Governance must ensure that communication remains coordinated, responsibilities remain clear, and significant leadership decisions are documented throughout the incident.

Technical recovery alone is insufficient.

Leadership must also preserve public confidence.

Applying the Defensible Evidence Framework™

The Defensible Evidence Framework™ provides government organizations with a practical model for demonstrating responsible stewardship before, during, and after cyber incidents.

Evidence should naturally exist throughout each governance domain.

Governance

- Agency cybersecurity governance policies
- Executive directives
- Governance committee charters
- Delegations of authority
- Strategic cybersecurity plans

Risk

- Enterprise risk assessments
- Critical services analysis
- Risk registers
- Risk acceptance documentation
- Third-party risk reviews

Oversight

- Governing body minutes
- Executive briefings
- Governance dashboards
- Budget oversight documentation

- Corrective action tracking

Operational

- Incident response plans
- Continuity of Operations (COOP) plans
- Disaster recovery testing
- Tabletop exercise reports
- Emergency communications plans

Assurance

- Internal audit reports
- Inspector General reviews
- Independent assessments
- Compliance audits
- Corrective action verification
- Governance maturity assessments

Collectively, these artifacts demonstrate that cybersecurity governance is integrated into the agency's broader responsibility for public stewardship.

Government Governance Checklist

Agency leadership should periodically confirm that they can answer "yes" to the following questions:

- Is cybersecurity governed as an enterprise and mission risk?
- Are governance responsibilities clearly assigned?

- Does executive leadership receive meaningful cybersecurity governance reporting?
- Are governing bodies regularly briefed on cyber risk?
- Are critical public services identified and prioritized?
- Are incident governance procedures regularly exercised?
- Are grant-related cybersecurity governance requirements documented?
- Can governance evidence be produced promptly during audits, legislative reviews, or investigations?
- Does independent assurance routinely evaluate cybersecurity governance?
- Are governance improvements documented and tracked to completion?

Governance Reality

Government organizations exist to serve the public.

Citizens rightly expect those services to remain available, resilient, and trustworthy—even during cyber incidents.

Technology enables government services.

Governance ensures those services can be protected, sustained, and defended.

When a significant cyber incident occurs, investigators, auditors, oversight bodies, legislators, and citizens will inevitably ask how leadership prepared for the event, how decisions were made, and whether public resources were managed responsibly.

. . .

The strongest answer is never a verbal explanation.

It is a well-organized body of evidence demonstrating that leadership exercised thoughtful, transparent, and accountable cybersecurity governance long before the incident occurred.

In government, good cybersecurity governance protects more than information systems.

It protects public confidence in the institutions entrusted to serve society.

Chapter 26
Nonprofit Organizations

Nonprofit organizations often assume they are unlikely targets for cyberattacks because they do not operate for profit.

The evidence tells a different story.

Charities, foundations, faith-based organizations, educational institutions, advocacy groups, humanitarian organizations, and social service agencies are increasingly targeted by ransomware operators, business email compromise schemes, phishing campaigns, and supply chain attacks. Attackers recognize that nonprofits frequently operate with limited budgets, small technology staffs, aging infrastructure, and extensive reliance on volunteers and third-party service providers.

Yet the greatest asset a nonprofit possesses is not its technology.

It is trust.

Donors trust that contributions will be used responsibly.

. . .

Beneficiaries trust that essential services will remain available.

Grantors trust that public and private funds are properly managed.

Volunteers trust that the organization protects the information they provide.

Cybersecurity governance protects that trust.

Governance Supports the Mission

Unlike commercial organizations, nonprofits exist to fulfill a mission.

Every cybersecurity decision should ultimately support that mission.

A cyber incident affecting a food bank may interrupt meal distribution.

A compromise at a domestic violence shelter may expose confidential client information.

A ransomware attack against a children's advocacy organization may delay critical services.

An attack on a faith-based organization may disrupt ministry, education, and community outreach.

. . .

Cybersecurity governance therefore protects more than information systems.

It protects the organization's ability to serve the people who depend upon it.

Boards Have Fiduciary Responsibilities

Many nonprofit Boards consist of dedicated volunteers who contribute expertise in finance, law, fundraising, healthcare, education, business, or community leadership.

Few join because they possess cybersecurity expertise.

Fortunately, cybersecurity expertise is not the Board's responsibility.

Governance is.

Directors should ensure that:

- cybersecurity is treated as an enterprise risk;
- executive leadership is accountable for cybersecurity;
- significant cyber risks are regularly reviewed;
- adequate resources are allocated;
- independent assurance is obtained when appropriate; and
- governance activities are documented.

. . .

Volunteer status does not reduce fiduciary responsibility.

Good governance is expected regardless of organizational size.

Limited Resources Require Better Governance

Nonprofits rarely have unlimited cybersecurity budgets.

That reality makes governance even more important.

Strong governance helps leadership prioritize limited resources according to organizational risk.

Boards should ask:

Which services are mission critical?

What information is most sensitive?

Which systems would most affect our mission if unavailable?

Where should limited cybersecurity funding have the greatest impact?

. . .

Governance enables informed prioritization rather than reactive spending.

Protecting Donor Confidence

Donors increasingly evaluate how organizations manage information.

A significant cyber incident can affect:

- donor confidence;
- fundraising campaigns;
- grant opportunities;
- community partnerships;
- corporate sponsorships; and
- long-term reputation.

Boards should recognize that cybersecurity governance contributes directly to organizational credibility.

Responsible stewardship includes protecting both financial contributions and the information entrusted to the organization.

Grant Funding and Accountability

Many nonprofits receive government grants, foundation funding, or other restricted financial support.

. . .

These funding sources increasingly expect organizations to demonstrate sound governance, internal controls, and cybersecurity practices.

Leadership should be prepared to demonstrate:

- risk management activities;
- governance oversight;
- internal controls;
- vendor management;
- incident response planning;
- corrective action tracking; and
- independent review where appropriate.

The Defensible Evidence Framework™ helps organizations organize this evidence before funding agencies request it.

Third-Party Dependence

Modern nonprofits often rely heavily on external providers.

These may include:

- cloud productivity platforms;
- donor management systems;
- fundraising applications;
- accounting services;
- managed IT providers;
- volunteer management platforms;

- payment processors; and
- software-as-a-service (SaaS) applications.

A weakness affecting one provider may affect the entire organization.

Boards should ensure that third-party risk becomes part of normal governance discussions rather than an afterthought during procurement.

Incident Governance

A cyber incident affecting a nonprofit requires leadership decisions extending beyond technical recovery.

Leadership may need to determine:

- how services will continue;
- when donors should be informed;
- whether grant agencies require notification;
- how volunteers should be coordinated;
- what legal obligations exist;
- how public communications will be managed; and
- how limited resources should be allocated during recovery.

These are governance decisions.

. . .

They should be documented as they occur.

Decision logs, executive briefings, Board communications, and after-action reviews become valuable evidence demonstrating responsible leadership.

Applying the Defensible Evidence Framework™

The Defensible Evidence Framework™ provides nonprofit organizations with a practical approach to governing cybersecurity while respecting limited resources.

Evidence should naturally exist across each governance domain.

Governance

- Board cybersecurity policies
- Board and committee minutes
- Executive accountability assignments
- Governance calendars
- Strategic technology plans

Risk

- Organizational risk assessments
- Mission-critical service analysis
- Risk registers
- Third-party risk reviews
- Risk acceptance documentation

Oversight

- Board dashboards
- Executive reports
- Corrective action tracking
- Governance committee reviews
- Annual Board education

Operational

- Incident response plans
- Business continuity plans
- Tabletop exercise reports
- Volunteer security awareness
- Vendor oversight documentation

Assurance

- Financial and internal audits
- Independent cybersecurity assessments
- Governance maturity assessments
- Evidence Gap Analyses
- Corrective action verification

These artifacts collectively demonstrate that cybersecurity governance supports the organization's mission and fulfills its stewardship responsibilities.

Nonprofit Governance Checklist

Boards should periodically confirm that they can answer "yes" to the following questions:

- Is cybersecurity recognized as a mission risk rather than solely an IT issue?
- Does the Board receive regular cybersecurity governance updates?
- Have mission-critical services been identified and prioritized?
- Are donor and beneficiary information appropriately protected?
- Are third-party service providers evaluated for cyber risk?
- Are cyber incidents included in organizational continuity planning?
- Are governance decisions documented as they occur?
- Can governance evidence be produced for grant reviews, audits, or regulatory inquiries?
- Does the organization periodically evaluate its cybersecurity governance maturity?
- Are governance improvements tracked through completion?

Governance Reality

Nonprofits exist because society depends upon the missions they serve.

Cyberattacks threaten more than computers.

. . .

They threaten meals delivered to families in need.

Advocacy for vulnerable populations.

Shelter for those in crisis.

Educational opportunities.

Medical services.

Community outreach.

Hope.

Cybersecurity governance helps ensure those missions continue despite an increasingly hostile threat landscape.

The strongest nonprofit organizations are not those with the largest technology budgets.

They are the ones whose leaders consistently demonstrate thoughtful stewardship, responsible oversight, and disciplined decision-making through evidence that reflects their commitment to the people and communities they serve.

. . .

For nonprofit organizations, cybersecurity governance is ultimately another expression of faithful stewardship.

And stewardship, when properly documented, becomes defensible evidence.

Chapter 27
Manufacturing

Manufacturing has become one of the most frequently targeted sectors for cyberattacks. Modern manufacturers rely on deeply interconnected information technology (IT), operational technology (OT), industrial control systems (ICS), robotics, programmable logic controllers (PLCs), cloud platforms, and global supply chains to produce goods efficiently and safely.

As these systems become more connected, cyber risk increasingly becomes operational risk.

A successful cyberattack can halt production, interrupt logistics, damage equipment, delay shipments, expose intellectual property, and threaten worker safety.

For manufacturing organizations, cybersecurity governance is no longer about protecting computers.

It is about protecting production.

Governance Protects the Production Mission

Every manufacturer has one overriding objective:

Keep production running safely, efficiently, and predictably.

Cybersecurity governance directly supports that objective.

A ransomware attack affecting production scheduling may idle multiple plants.

A compromise involving industrial control systems may interrupt manufacturing lines.

A supply chain attack may delay raw materials or introduce malicious software into production environments.

The Board's responsibility is to ensure these risks are governed as enterprise risks rather than isolated technology problems.

Information Technology and Operational Technology

Manufacturing organizations operate two interconnected technology environments.

Information Technology (IT) supports business operations.

. . .

Operational Technology (OT) controls physical production.

Examples include:

Information Technology

- Enterprise Resource Planning (ERP)
- Email
- Financial systems
- Engineering documentation
- Human resources
- Customer relationship management

Operational Technology

- Industrial Control Systems (ICS)
- Programmable Logic Controllers (PLCs)
- Supervisory Control and Data Acquisition (SCADA)
- Manufacturing Execution Systems (MES)
- Robotics
- Process automation
- Plant monitoring systems

Governance must bridge both environments.

Cybersecurity decisions affecting one frequently influence the other.

Production Downtime Is Enterprise Risk

Manufacturing organizations often calculate cyber losses differently from other industries.

The greatest financial impact frequently comes from production downtime.

Every hour of interrupted manufacturing may result in:

- lost revenue;
- delayed customer deliveries;
- contractual penalties;
- supply chain disruption;
- overtime expenses;
- quality assurance delays; and
- reputational damage.

Boards should therefore evaluate cyber risk using operational business metrics rather than purely technical measurements.

Intellectual Property Requires Governance

Manufacturing organizations often possess valuable intellectual property.

Examples include:

- product designs;

- engineering drawings;
- formulas;
- production methods;
- research data;
- quality processes;
- proprietary software; and
- supplier specifications.

The theft of intellectual property may affect competitive advantage for years.

Governance should ensure these assets receive appropriate protection and executive oversight.

Supply Chain Risk

Manufacturers rarely operate independently.

Production depends upon an ecosystem of suppliers, logistics providers, distributors, contractors, and technology vendors.

Cyber incidents affecting a single supplier may interrupt multiple production facilities.

Boards should require governance processes addressing:

- supplier cybersecurity assessments;
- critical supplier identification;

- contractual cybersecurity requirements;
- continuous monitoring;
- contingency planning; and
- alternative sourcing strategies.

Supply chain resilience has become a governance responsibility.

Safety and Cybersecurity

Many manufacturing environments involve physical hazards.

Cyber incidents affecting operational technology may influence:

- worker safety;
- environmental controls;
- hazardous materials;
- equipment operation;
- quality assurance; and
- regulatory compliance.

Cybersecurity governance should therefore coordinate closely with safety governance.

Leadership should understand that operational disruption may create consequences extending well beyond financial loss.

Incident Governance

Manufacturing cyber incidents often require rapid executive decision-making.

Leadership may need to determine:

- whether production should stop;
- which facilities should continue operating;
- whether manual operations are feasible;
- how customers will be informed;
- whether suppliers require notification;
- how recovery priorities should be established; and
- how contractual obligations will be managed.

These decisions require governance.

Decision logs, executive briefings, Board communications, and recovery prioritization records become critical evidence demonstrating thoughtful leadership.

Applying the Defensible Evidence Framework™

The Defensible Evidence Framework™ enables manufacturers to demonstrate disciplined governance across both business and production operations.

Evidence should naturally exist throughout each governance domain.

Governance

- Board cybersecurity oversight
- Executive governance committees
- Cybersecurity governance policies
- Operational technology governance policies
- Strategic resilience plans

Risk

- Enterprise cyber risk assessments
- Operational technology risk assessments
- Critical production asset inventories
- Supply chain risk assessments
- Risk acceptance documentation

Oversight

- Board dashboards
- Executive manufacturing risk reports
- Production resilience reviews
- Corrective action tracking
- Governance committee minutes

Operational

- Incident response plans
- Operational technology recovery plans
- Business continuity exercises
- Tabletop exercise reports
- Production recovery procedures

Assurance

- Internal audit reports
- Operational technology assessments
- Independent cybersecurity reviews
- Penetration testing summaries
- Governance maturity assessments

Together, these artifacts demonstrate that cybersecurity governance extends across the entire manufacturing enterprise.

Manufacturing Governance Checklist

Boards should periodically confirm that they can answer "yes" to the following questions:

- Is operational technology included in enterprise cyber governance?
- Have critical production systems been identified and prioritized?
- Are cybersecurity risks evaluated according to their operational impact?
- Are intellectual property risks regularly reviewed?
- Are critical suppliers evaluated for cybersecurity resilience?
- Are production recovery procedures regularly exercised?
- Does executive leadership report on manufacturing cyber resilience rather than only IT security?
- Are governance decisions documented throughout significant cyber events?

- Does independent assurance evaluate both IT and operational technology governance?
- Can governance evidence be produced quickly following a cyber incident?

Governance Reality

Manufacturing organizations compete on reliability.

Customers expect products to be delivered safely, on time, and without interruption.

Cybersecurity governance helps make that possible.

Technology protects machinery.

Operational controls protect production.

Governance protects the organization that depends on both.

When a significant cyber incident affects a manufacturer, the lasting question is rarely whether the production line stopped.

The lasting question is whether leadership anticipated the risk, governed it responsibly, and can demonstrate those decisions through credible evidence.

. . .

Manufacturers that adopt evidence-driven governance strengthen more than cybersecurity.

They strengthen operational resilience, supply chain confidence, customer trust, and long-term competitiveness.

In manufacturing, good governance keeps more than systems running.

It keeps the business moving.

Part VIII

Implementing the Framework

Chapter 28
The 90-Day Roadmap

Knowing what good governance looks like is valuable.

Building it is transformational.

Throughout this book, we have explored the principles of the Defensible Evidence Framework™. We have examined governance evidence, risk evidence, oversight evidence, operational evidence, assurance evidence, evidence lifecycle management, governance maturity, evidence gap analysis, and incident governance.

The obvious question now becomes:

Where do we begin?

The answer is not to rewrite every policy.

It is not to purchase new governance software.

. . .

It is not to create hundreds of new documents.

Instead, organizations should begin by establishing a disciplined governance foundation that produces evidence naturally.

This chapter provides a practical ninety-day roadmap for doing exactly that.

The roadmap is intentionally realistic.

It assumes the organization already has governance activities underway.

The objective is to strengthen them—not replace them.

Guiding Principles

Before discussing specific activities, three principles should guide every implementation.

First, improve governance before improving documentation.

Better governance naturally produces better evidence.

Second, build upon existing governance processes whenever possible.

. . .

Do not create unnecessary bureaucracy.

Third, pursue continuous improvement rather than perfection.

Governance maturity develops over time.

The objective is progress.

Days 1–30: Understand Your Current State

The first month is devoted to discovery.

Leadership cannot improve governance without understanding its current condition.

Begin by identifying every significant governance activity already taking place.

Ask questions such as:

How does the Board oversee cybersecurity?

How is cyber risk reported?

Who approves governance policies?

. . .

Who accepts significant cyber risk?

How are corrective actions tracked?

Who performs independent assurance?

Just as importantly, determine where governance evidence currently resides.

Many organizations discover evidence spread across:

Board portals.

Document management systems.

SharePoint sites.

Email archives.

Audit repositories.

Risk management applications.

. . .

Departmental file shares.

Personal folders.

The objective during the first thirty days is not to reorganize anything.

It is simply to understand what already exists.

Perform an Initial Evidence Gap Analysis

Using the five evidence domains as a guide, evaluate whether governance evidence can be readily produced.

Review:

Governance Evidence.

Risk Evidence.

Oversight Evidence.

Operational Evidence.

. . .

Assurance Evidence.

Document missing artifacts.

Identify inconsistent practices.

Note areas where evidence exists but cannot easily be located.

The purpose is clarity.

Not criticism.

Every organization begins with gaps.

Knowing where they exist is the first step toward closing them.

Identify Governance Owners

Every evidence domain should have clearly identified ownership.

Ownership does not necessarily mean creating every document.

It means ensuring that governance activities occur and that evidence is preserved appropriately.

. . .

Typical ownership may include:

Corporate Governance.

Enterprise Risk Management.

Internal Audit.

Legal.

Compliance.

Information Security.

Executive leadership.

Regardless of organizational structure, accountability should always be explicit.

Unassigned governance responsibilities rarely remain effective over time.

Days 31–60: Strengthen Governance

The second month focuses on improving governance itself.

Review governance authority.

Update committee charters where necessary.

Clarify reporting relationships.

Formalize risk acceptance procedures.

Review Board reporting.

Confirm executive accountability.

Update governance calendars.

Improve escalation procedures.

Most importantly, improve the quality of governance conversations.

Board reporting should evolve beyond technical metrics.

. . .

Executives should receive decision-oriented information.

Committees should document not only presentations but discussions, questions, and follow-up actions.

Governance becomes stronger when leadership becomes more engaged.

Establish the Governance Evidence Repository

During the second month, organizations should also establish their Governance Evidence Repository.

The repository does not require specialized software.

Existing enterprise platforms often provide everything required.

Organize the repository around the five evidence domains.

Develop naming conventions.

Assign ownership.

Define metadata standards.

. . .

Establish version control.

Implement access controls.

Document retention requirements.

The objective is creating a governance library that preserves institutional knowledge rather than simply storing documents.

Standardize Governance Evidence

Consistency strengthens credibility.

Develop standard templates for common governance activities.

Examples include:

Board cybersecurity reports.

Risk acceptance decisions.

Committee minutes.

Governance dashboards.

. . .

Tabletop exercise reports.

After-action reviews.

Corrective action tracking.

Evidence inventories.

Standardization reduces ambiguity.

It also makes governance evidence easier to review, compare, and preserve over time.

Days 61–90: Validate the Framework

The final month focuses on testing governance.

Not through another audit.

Through practical validation.

Conduct a tabletop exercise emphasizing governance rather than technical response.

Evaluate:

. . .

Board notification.

Executive decision-making.

Escalation.

Communications.

Evidence preservation.

Decision logging.

Risk acceptance.

After the exercise, ask participants:

Could we demonstrate every significant governance decision we made?

If the answer is uncertain, identify why.

. . .

The objective is to discover weaknesses while the environment remains controlled.

Test the Repository

Choose several governance questions at random.

For example:

Produce every Board discussion concerning ransomware over the last two years.

Locate all risk acceptance decisions involving cloud services.

Find every internal audit finding related to cybersecurity governance.

Identify all completed tabletop exercises.

Measure the time required to locate each artifact.

If evidence cannot be produced quickly, governance maturity still has room to improve.

Accessibility is a governance capability.

Conduct a Readiness Assessment

At the conclusion of ninety days, complete the Governance Readiness Scorecard.

Evaluate each evidence domain objectively.

Determine your current maturity level.

Reactive.

Developing.

Managed.

Measured.

Defensible.

Do not treat the score as a final judgment.

Treat it as a baseline.

. . .

Every future assessment should demonstrate measurable improvement.

Celebrate Progress

Governance programs often focus exclusively on deficiencies.

That approach discourages leadership.

Instead, acknowledge improvements.

Perhaps Board reporting has become more strategic.

Perhaps risk ownership is now clearly defined.

Perhaps evidence is easier to locate.

Perhaps corrective actions are tracked consistently.

These improvements deserve recognition.

Governance maturity is built one improvement at a time.

What Happens After Ninety Days?

Ninety days does not complete governance.

It establishes momentum.

Following implementation, organizations should transition into a continuous governance cycle.

Quarterly activities may include:

- Governance committee reviews.
- Governance Evidence Repository updates.
- Evidence Gap Analyses.
- Board cybersecurity briefings.
- Risk register reviews.
- Corrective action status reviews.
- Governance metrics reporting.

Annual activities should include:

- Governance Readiness Scorecard reassessment.
- Independent governance assurance.
- Board governance education.
- Governance policy review.
- Tabletop exercises emphasizing executive leadership and Board oversight.
- Strategic governance planning.

. . .

Governance maturity is achieved through consistency, not intensity.

Small Improvements, Significant Results

One of the most encouraging aspects of the Defensible Evidence Framework™ is that organizations rarely need to start from scratch.

Most already possess governance.

Most already produce evidence.

Most already have capable leaders.

The challenge is bringing these elements together into a disciplined, evidence-driven governance program.

Small improvements produce significant results.

A better Board discussion.

A more complete decision log.

A clearer risk acceptance process.

. . .

A stronger after-action review.

A more organized repository.

Over time, these incremental improvements transform governance.

The First Step

Organizations often delay governance improvement because they believe the effort will be overwhelming.

It rarely is.

The first step is surprisingly simple.

Schedule a meeting with the people responsible for governing cybersecurity.

Ask one question.

> **"If we needed to demonstrate reasonable cybersecurity governance today, could we?"**

The discussion that follows may become the most valuable governance conversation your organization has ever had.

Because every defensible organization begins the same way.

Not with technology.

Not with compliance.

Not with documentation.

But with leadership committed to governing cybersecurity intentionally—and preserving the evidence to prove it.

Chapter 29
The First Year

Governance is not transformed in ninety days.

It is transformed through the disciplined execution of hundreds of small decisions over time.

The ninety-day roadmap described in the previous chapter establishes direction. It creates governance structure, clarifies accountability, and begins organizing evidence. Those first steps are important.

They are not the destination.

The first year is where governance begins to mature.

Processes become habits.

Evidence becomes routine.

. . .

Leadership conversations become more strategic.

Governance shifts from being a project to becoming part of the organization's operating model.

This chapter outlines what organizations should expect during the first year of implementing the Defensible Evidence Framework™.

Quarter One: Building the Foundation

The first quarter focuses on establishing consistency.

Governance structures have been defined.

Evidence owners have been identified.

The Governance Evidence Repository has been established.

The Governance Readiness Scorecard has provided an initial baseline.

Now leadership begins operating within the framework.

Board reports become more consistent.

. . .

Committee responsibilities become clearer.

Risk discussions become more structured.

Evidence begins accumulating intentionally.

The objective during this quarter is not perfection.

It is repeatability.

Leadership should resist the temptation to launch dozens of governance initiatives simultaneously.

Instead, focus on making the new governance processes dependable.

Consistency is more valuable than complexity.

Quarter Two: Establishing Governance Rhythm

By the second quarter, governance should begin developing a predictable cadence.

Cybersecurity becomes a recurring topic at Board and committee meetings.

. . .

Risk registers receive scheduled reviews.

Governance dashboards mature.

Corrective actions are monitored.

Policies are reviewed according to established schedules.

Internal Audit incorporates governance into its planning.

The organization begins operating according to governance rather than reacting to individual events.

This rhythm is important because predictable governance naturally creates predictable evidence.

Leadership no longer asks whether governance activities occurred.

They know they occurred because governance follows an established calendar.

Quarter Three: Measuring Effectiveness

Once governance becomes routine, leadership should begin evaluating its effectiveness.

. . .

Are Board discussions producing better decisions?

Are executive reports supporting meaningful oversight?

Are risks being escalated appropriately?

Are corrective actions closing on time?

Can governance evidence be located quickly?

Has the quality of evidence improved?

This is where organizations begin moving beyond activity toward performance.

Governance itself becomes measurable.

Rather than asking, "Did we hold the meeting?"

Leadership begins asking, "Did the meeting improve governance?"

That subtle shift represents significant organizational maturity.

Quarter Four: Continuous Improvement

By the fourth quarter, governance should begin feeding itself.

Lessons from audits improve policies.

Lessons from tabletop exercises improve decision-making.

Lessons from Board discussions improve reporting.

Lessons from incidents improve preparedness.

Governance becomes a learning system.

The organization no longer waits for external pressure to improve.

Improvement becomes part of normal operations.

This is one of the defining characteristics of mature governance.

Organizations stop governing for compliance.

They begin governing for organizational resilience.

The Board's First Annual Review

At the end of the first year, the Board should conduct a formal review of cybersecurity governance.

Not merely cybersecurity.

Governance.

Leadership should revisit questions such as:

Have governance responsibilities become clearer?

Has Board oversight improved?

Has executive reporting become more useful?

Are governance decisions documented consistently?

Has independent assurance strengthened confidence?

What evidence gaps remain?

How has governance maturity improved?

. . .

This discussion should be strategic rather than operational.

It should focus on how leadership governs cybersecurity—not on the latest threat landscape.

Reassessing Governance Readiness

The Governance Readiness Scorecard should be repeated near the end of the first year.

Not to produce a better number.

To measure progress.

Organizations often discover that some evidence domains improve faster than others.

Governance Evidence may mature rapidly because structures become well defined.

Operational Evidence may improve more gradually as exercises, training, and recovery testing become routine.

Assurance Evidence often strengthens after Internal Audit incorporates governance into its annual work plan.

. . .

The reassessment helps leadership identify where additional attention is needed during the second year.

The Repository Becomes Institutional Memory

During the first year, the Governance Evidence Repository begins demonstrating its true value.

Leadership changes may occur.

Policies may be revised.

Committee memberships may evolve.

Technology platforms may change.

The repository preserves continuity.

It becomes the single source of truth for governance evidence.

Rather than depending upon institutional memory, the organization begins depending upon documented governance history.

. . .

This transition significantly strengthens organizational resilience.

Developing Better Questions

One of the clearest signs of governance maturity is the evolution of leadership questions.

Early in implementation, leaders ask:

Do we have a policy?

Did we complete the assessment?

Did we conduct the exercise?

Later, those questions become:

Did the policy improve governance?

Did the assessment influence decisions?

Did the exercise strengthen preparedness?

Can we demonstrate what we learned?

. . .

Eventually, leadership asks perhaps the most important governance question of all:

What evidence demonstrates that our governance is improving?

That question marks a profound shift in organizational thinking.

Common First-Year Challenges

Nearly every organization encounters similar challenges during the first year.

Competing priorities delay governance initiatives.

Leadership turnover interrupts momentum.

Evidence ownership remains unclear.

Legacy documentation requires organization.

Governance activities become inconsistent during busy operational periods.

. . .

These challenges should be expected.

Governance maturity does not develop without resistance.

Organizations that acknowledge these realities are far more likely to sustain long-term improvement.

What Success Looks Like

Success after the first year should not be measured by the number of policies written or the volume of evidence collected.

Instead, leadership should observe meaningful changes.

Board members ask more informed questions.

Executives receive clearer risk information.

Decision authority is understood.

Evidence is easier to locate.

Corrective actions remain visible until completed.

Internal Audit evaluates governance with greater confidence.

. . .

Governance discussions become part of ordinary business.

These changes indicate that governance is becoming embedded within the organization.

Looking Ahead

At the conclusion of the first year, organizations often recognize something unexpected.

The Defensible Evidence Framework™ did not create entirely new governance.

It improved the governance they already possessed.

Leadership became more intentional.

Evidence became more disciplined.

Oversight became more visible.

Decision-making became more transparent.

Governance became more defensible.

. . .

That realization prepares the organization for the next stage of its journey.

No longer asking how to implement governance...

...but how to sustain it for years to come.

Because truly mature governance is not measured by what an organization accomplishes during its first year.

It is measured by what leadership continues doing in every year that follows.

Chapter 30
Creating a Culture of Evidence

Technology changes.

Processes evolve.

Leadership turns over.

Culture endures.

Every organization has a culture.

Some cultivate innovation.

Others emphasize operational excellence.

Some value customer service.

Others prioritize financial discipline.

. . .

Whether intentionally or unintentionally, every organization also develops a governance culture.

The question is whether that culture produces evidence.

A culture of evidence is not one in which employees create more documentation.

It is one in which people naturally preserve the decisions, accountability, and oversight that define responsible leadership.

That distinction is fundamental.

The Defensible Evidence Framework™ is not ultimately about documents.

It is about organizational behavior.

Culture Shapes Governance

Policies establish expectations.

Governance defines responsibilities.

. . .

Culture determines whether either actually happens.

Two organizations can operate under identical governance structures while producing dramatically different results.

In one organization, leaders document decisions because governance requires it.

In another, leaders document decisions because they understand the value of preserving organizational memory.

The first organization complies.

The second governs.

The difference is culture.

Evidence Is a Leadership Value

Cultures begin with leadership.

Employees quickly recognize what leadership values.

If executives consistently ask:

. . .

"Was this documented?"

"Who approved this?"

"What evidence supports that decision?"

"Has the Board been informed?"

Then evidence becomes part of everyday leadership.

Conversely, when leaders rely on verbal agreements, undocumented decisions, and institutional memory, employees learn that documentation is optional.

Culture follows behavior.

Leadership establishes the example.

Documentation Is Not Bureaucracy

One of the greatest obstacles to building a culture of evidence is the belief that documentation creates unnecessary bureaucracy.

Poor documentation certainly can.

. . .

Meaningful evidence does not.

Consider a documented risk acceptance decision.

Its purpose is not administrative.

Its purpose is to preserve:

The decision.

The rationale.

The alternatives considered.

The authority exercised.

The accountability accepted.

Years later, that single document may explain a decision no one remembers making.

Evidence preserves organizational knowledge.

. . .

It is not paperwork.

It is institutional memory.

Every Employee Creates Evidence

Many leaders assume governance evidence belongs exclusively to executives and Boards.

It does not.

Governance evidence is created throughout the organization.

Managers document approvals.

Project leaders document decisions.

Security teams document investigations.

Risk managers document assessments.

Auditors document findings.

Procurement documents vendor reviews.

. . .

Legal documents governance obligations.

Every employee contributes another piece of the governance story.

Creating a culture of evidence means helping every employee understand that the records they create today may become critical governance evidence tomorrow.

Reward the Right Behaviors

Organizations reinforce whatever they recognize.

If leadership celebrates only rapid execution, employees naturally prioritize speed.

If leadership values thoughtful decision-making, accountability, and documentation, those behaviors become cultural norms.

A culture of evidence should recognize behaviors such as:

Escalating concerns promptly.

Documenting significant decisions.

Completing corrective actions.

. . .

Updating governance records.

Sharing lessons learned.

Preserving institutional knowledge.

Improving governance processes.

These actions rarely attract attention.

Collectively, they define governance maturity.

Normalize Governance Conversations

Evidence should never become something discussed only during audits or after cyber incidents.

Organizations with mature governance discuss evidence routinely.

Board meetings include governance discussions.

Executive meetings review decision logs.

Project reviews examine governance implications.

. . .

Risk discussions consider evidence quality.

After-action reviews evaluate documentation alongside operational performance.

When evidence becomes part of normal conversation, governance becomes part of normal business.

Remove the Fear

Some organizations unintentionally create resistance to documentation.

Employees worry that written records will later be used to assign blame.

As a result, decisions remain undocumented.

Important discussions occur verbally.

Emails replace formal records.

Governance weakens.

A culture of evidence takes a different approach.

. . .

Evidence exists to demonstrate responsible leadership.

Not to punish responsible decision-makers.

Organizations should encourage honest documentation, thoughtful discussion, and transparent decision-making.

Leaders who fear documentation often create greater risk than the documentation itself.

Build Governance into Everyday Work

The strongest evidence is produced without anyone consciously trying to create evidence.

Board packets become routine.

Meeting minutes follow standard formats.

Risk reviews update the enterprise risk register automatically.

Corrective actions remain visible until closure.

Policy reviews occur according to established schedules.

. . .

Evidence becomes the natural output of governance.

This is the hallmark of organizational maturity.

Evidence no longer requires extraordinary effort.

It simply reflects how the organization operates.

Make Evidence Accessible

Creating evidence is only half the challenge.

People must also know how to find it.

Employees should understand:

Where governance records belong.

Who owns them.

How they are classified.

How they are protected.

. . .

How long they are retained.

How they support leadership.

Accessibility strengthens confidence.

If employees cannot locate governance evidence, they eventually stop relying upon it.

Culture requires both creation and stewardship.

Teach the "Why"

Organizations frequently teach employees *what* to document.

Far fewer explain *why*.

People are far more likely to preserve meaningful evidence when they understand its purpose.

Explain that governance evidence:

Protects leadership.

. . .

Supports transparency.

Improves decision-making.

Preserves institutional knowledge.

Strengthens organizational resilience.

Demonstrates accountability.

Builds stakeholder trust.

When people understand the purpose behind documentation, compliance evolves into commitment.

Culture Is Measured During Crisis

The true test of a culture of evidence occurs during uncertainty.

When a significant cyber incident unfolds, does documentation continue?

Do leaders maintain decision logs?

Are governance discussions preserved?

. . .

Does accountability remain clear?

Or does the organization abandon discipline under pressure?

Cultures that value evidence become stronger during crises.

They understand that today's governance decisions will become tomorrow's governance record.

Pressure reinforces their discipline rather than replacing it.

The Legacy of Leadership

Every leadership team leaves a legacy.

Sometimes it is measured in financial performance.

Sometimes in organizational growth.

Sometimes in innovation.

Increasingly, it is measured in governance.

. . .

Future Boards will inherit today's governance structures.

Future executives will rely on today's decision records.

Future auditors will review today's oversight.

Future investigators may examine today's governance evidence.

Leaders therefore create more than decisions.

They create the historical record by which those decisions will someday be judged.

That record deserves intentional stewardship.

A Culture That Sustains Itself

The highest expression of governance maturity occurs when evidence becomes self-sustaining.

Employees no longer document decisions because someone reminds them.

Boards no longer request governance evidence because they expect it.

. . .

Executives no longer ask whether governance occurred because the evidence already exists.

Governance becomes embedded in organizational identity.

Evidence becomes part of organizational culture.

At that point, the Defensible Evidence Framework™ is no longer simply a framework.

It becomes the way the organization leads.

The Defensible Organization

Throughout this book, one central idea has remained unchanged.

The strongest governance evidence is created naturally as governance occurs—not reconstructed months later.

Creating a culture of evidence transforms that principle from a governance objective into an organizational habit.

It ensures that every meaningful decision, every act of oversight, every assessment of risk, and every exercise of leadership contributes to a growing body of evidence that reflects how the organization governs itself.

. . .

Technology will continue to evolve.

Threats will continue to emerge.

Regulations will continue to change.

But organizations that build a culture of evidence will possess something far more enduring:

A culture in which governance is visible, accountability is expected, and defensibility is simply the natural outcome of responsible leadership.

The Governance That Endures

Every generation of leadership inherits a different set of challenges.

For one generation, it was industrialization.

For another, globalization.

Then came digital transformation.

Today, cybersecurity has joined that list—not as an information technology problem, but as one of the defining governance responsibilities of modern leadership.

Cybersecurity will continue to evolve.

Artificial intelligence will introduce new opportunities and new risks.

Quantum computing will reshape cryptography.

Supply chains will become more interconnected.

Regulations will become more demanding.

Threat actors will become more sophisticated.

Technology will change.

The principles of good governance will not.

Throughout this book, we have examined governance from a different perspective than most cybersecurity literature.

We have not focused on stopping attackers.

We have not concentrated on security controls.

We have not attempted to predict the next threat.

Instead, we have explored a more enduring question:

> **How can leadership demonstrate that it governed cybersecurity responsibly?**

That question matters because every organization will eventually face scrutiny.

Perhaps from regulators.

Perhaps from insurers.

Perhaps from auditors.

Perhaps from shareholders.

Perhaps from customers.

Perhaps from courts.

Perhaps from its own Board.

The audience will differ.

The questions will not.

The Shift from Compliance to Confidence

Many organizations still approach cybersecurity governance as a compliance exercise.

Comply with regulations.

Comply with standards.

Comply with contractual obligations.

Compliance remains important.

It is not sufficient.

Compliance demonstrates that requirements were met.

Governance demonstrates that leadership exercised judgment.

The Defensible Evidence Framework™ represents a shift away from governance driven primarily by compliance toward governance driven by confidence.

Confidence that leadership understands organizational risk.

Confidence that oversight is functioning.

Confidence that accountability is clear.

Confidence that important decisions are documented.

Confidence that governance can withstand independent scrutiny.

Confidence ultimately comes from evidence.

Evidence Is the Language of Leadership

Every profession relies upon evidence.

Physicians rely upon medical records.

Engineers rely upon design documentation.

Attorneys rely upon case law.

Accountants rely upon financial records.

Governance is no different.

Leadership communicates through evidence.

Board minutes communicate oversight.

Policies communicate expectations.

Risk registers communicate priorities.

Decision logs communicate judgment.

Audit reports communicate assurance.

Together, they tell the story of leadership.

Without evidence, governance becomes opinion.

With evidence, governance becomes demonstrable.

The Legacy You Leave

Boards change.

Chief executive officers retire.

Chief information security officers accept new opportunities.

Committee memberships evolve.

Technology platforms disappear.

What remains is the governance record.

Years from now, people who never met today's leadership may review the decisions being made today.

They will never hear the discussions.

They will never witness the debates.

They will never experience the uncertainty.

They will know leadership only through its evidence.

That realization changes how we should think about governance.

Every Board packet becomes part of the organization's history.

Every governance decision becomes part of its institutional memory.

Every documented lesson becomes part of its resilience.

Evidence preserves leadership long after leadership has changed.

Governance Is an Investment

Organizations often ask whether governance is worth the effort.

The better question is this:

What is the cost of governing without evidence?

The answer rarely appears on a balance sheet.

It appears during investigations.

Insurance reviews.

Regulatory examinations.

Shareholder lawsuits.

Board inquiries.

Executive transitions.

Major cyber incidents.

In those moments, governance either creates confidence or creates uncertainty.

Evidence is the difference.

Viewed through that lens, governance is not overhead.

It is one of the organization's most valuable long-term investments.

The Defensible Organization

Early in this book, we introduced the Monday Morning Question.

If an investigator asked for evidence of your cybersecurity governance tomorrow morning...

What would you produce?

Now, having completed the journey through the Defensible Evidence Framework™, that question should feel different.

The objective is no longer to search for documentation.

The objective is to build an organization where the answer already exists.

A defensible organization is not one that never experiences cyber incidents.

Such organizations do not exist.

A defensible organization is one that consistently demonstrates:

Clear governance.

Thoughtful leadership.

Disciplined oversight.

Deliberate decision-making.

Continuous improvement.

Credible evidence.

That is an achievable goal.

Not because technology makes it possible.

Because leadership chooses to make it so.

The Framework Is Only the Beginning

If this book has accomplished its purpose, it has done more than introduce a new governance framework.

It has encouraged a different way of thinking.

You may begin asking different questions at your next Board meeting.

You may redesign executive reporting.

You may reorganize governance evidence.

You may rethink how risk acceptance is documented.

You may establish a Governance Evidence Repository.

You may conduct your first Evidence Gap Analysis.

You may launch your first Governance Readiness Assessment.

Those are valuable outcomes.

But the greatest outcome would be something even simpler.

That leadership begins asking, consistently:

> **"What evidence should naturally exist if we are governing well?"**

That single question has the power to transform governance.

A Personal Reflection

When I began developing the Defensible Evidence Framework™, I was not trying to create another cybersecurity framework.

The profession already has exceptional technical frameworks.

What I saw, repeatedly, was something different.

Organizations with capable leaders.

Talented security professionals.

Reasonable governance.

Yet when difficult questions were asked, they struggled to demonstrate what they had done.

Not because governance was absent.

Because evidence was.

That observation became the foundation for this work.

The strongest governance evidence is created naturally as governance occurs—not reconstructed months later.

Everything in this book flows from that single principle.

If organizations embrace it, governance changes.

Not because they create more documentation.

But because they lead more intentionally.

The Final Question

Every book should leave the reader with one idea worth remembering.

For this book, it is the same question with which we began.

Not because it is easy.

Because it is enduring.

Tomorrow morning...

If your Board asked...

If your regulator asked...

If your insurer asked...

If your customers asked...

If a court asked...

If history asked...

> **Could you demonstrate that your organization exercised reasonable cybersecurity governance?**

If the answer is yes, your governance is becoming defensible.

If the answer is no, your work begins today.

And that is precisely why this framework exists.

A Final Thought

Cybersecurity is ultimately about protecting systems.

Governance is ultimately about protecting organizations.

Evidence is ultimately about protecting leadership.

When those three disciplines work together, something remarkable happens.

Governance becomes visible.

Accountability becomes credible.

Leadership becomes defensible.

That is the promise of the **Defensible Evidence Framework™**.

And that is the future of cybersecurity governance.

Appendix A—Governance Evidence Catalog

Purpose

The Defensible Evidence Framework™ identifies five evidence domains that collectively demonstrate reasonable cybersecurity governance. This appendix serves as a practical reference catalog of governance artifacts that organizations should consider maintaining. It is not intended to be a mandatory checklist. Rather, it provides examples of the types of evidence that naturally result from disciplined governance.

Not every organization will require every artifact listed here. The appropriate evidence will vary based on organizational size, industry, regulatory obligations, and governance maturity.

The guiding principle remains consistent:

> **The strongest governance evidence is created naturally as governance occurs—not reconstructed after an incident.**

Domain 1 — Governance Evidence

Governance Evidence demonstrates that leadership intentionally established cybersecurity governance.

Governance Structure

- Board Cybersecurity Governance Charter
- Board Committee Charters
- Executive Cybersecurity Governance Charter
- Cybersecurity Governance Policy
- Enterprise Governance Framework
- Governance Operating Model
- Governance Calendar
- Governance Committee Membership Lists
- Governance Organizational Charts
- Governance Delegations of Authority

Leadership Accountability

- Executive accountability assignments
- CISO reporting structure
- Cybersecurity RACI Matrix
- Board responsibility statements
- Committee responsibility matrix
- Governance decision authority matrix
- Executive succession documentation

Policy Governance

- Board-approved policies
- Policy approval records
- Policy review schedules
- Policy revision history
- Policy exception approvals
- Policy distribution records
- Policy acknowledgment tracking

Domain 2 – Risk Evidence

Risk Evidence demonstrates how leadership governs cybersecurity risk.

Enterprise Risk

- Enterprise Cyber Risk Register
- Risk Assessment Reports
- Business Impact Analyses
- Critical Asset Inventory
- Crown Jewel Analysis
- Risk Prioritization Matrix
- Risk Heat Maps
- Threat Scenario Analyses

Risk Decisions

- Risk Acceptance Forms
- Executive Risk Approval Records
- Residual Risk Documentation
- Risk Mitigation Plans
- Risk Transfer Decisions
- Insurance Risk Reviews
- Board Risk Reports

Third-Party Risk

- Vendor Risk Assessments
- Supplier Security Reviews
- Due Diligence Questionnaires
- Third-Party Monitoring Reports

- Contract Security Requirements
- Service-Level Reviews
- Vendor Exception Approvals

Domain 3 – Oversight Evidence

Oversight Evidence demonstrates active leadership engagement.

Board Oversight

- Board Meeting Minutes
- Board Cybersecurity Briefings
- Board Dashboards
- Director Education Records
- Board Action Registers
- Board Resolutions
- Committee Reports

Executive Oversight

- Executive Steering Committee Minutes
- Executive Dashboards
- Executive Decision Records
- Executive Risk Reviews
- Governance Status Reports
- Quarterly Governance Reports

Governance Monitoring

- Corrective Action Register
- Action Item Tracker
- Governance Metrics

- Governance KPIs
- Governance Scorecards
- Escalation Records
- Management Review Records

Domain 4 — Operational Evidence

Operational Evidence demonstrates that governance decisions became organizational actions.

Incident Preparedness

- Incident Response Plan
- Crisis Communications Plan
- Incident Escalation Procedures
- Incident Call Trees
- Decision Logs
- Situation Reports (SITREPs)
- Incident Status Reports

Exercises

- Tabletop Exercise Plans
- Participant Lists
- Facilitator Guides
- Exercise Results
- Lessons Learned Reports
- Improvement Plans
- Exercise Validation Reports

Business Resilience

- Business Continuity Plans
- Disaster Recovery Plans
- Recovery Test Results
- Backup Validation Reports
- Operational Recovery Metrics
- Recovery Readiness Assessments

Operational Governance

- Security Awareness Metrics
- Phishing Campaign Results
- Vulnerability Management Reports
- Patch Governance Reports
- Privileged Access Reviews
- Operational Risk Reviews
- Vendor Performance Reviews

Domain 5 — Assurance Evidence

Assurance Evidence demonstrates independent validation of governance.

Internal Assurance

- Internal Audit Reports
- Audit Plans
- Audit Work Programs
- Audit Recommendations
- Management Responses
- Follow-up Reviews

- Closure Verification

External Assurance

- External Audit Reports
- Independent Governance Assessments
- Cybersecurity Maturity Assessments
- ISO Assessment Reports
- Regulatory Examination Reports
- Independent Advisory Reports

Technical Assurance

- Penetration Test Reports
- Red Team Reports
- Purple Team Reports
- Vulnerability Validation Reports
- Security Architecture Reviews
- Configuration Assessments

Governance Validation

- Governance Readiness Assessments
- Evidence Gap Analyses
- Governance Maturity Assessments
- Governance Health Checks
- Independent Board Reviews
- Governance Benchmarking Reports

Cross-Domain Evidence

Certain governance artifacts support multiple evidence domains simultaneously.

Examples include:

- Board cybersecurity presentations
- Annual governance reports
- Enterprise risk reports
- Executive dashboards
- Governance maturity assessments
- Cyber insurance applications
- Regulatory submissions
- Governance committee minutes
- Post-incident governance reviews
- Annual cybersecurity strategy documents

These artifacts frequently become some of the strongest pieces of governance evidence because they demonstrate the interaction between leadership, risk, oversight, operations, and assurance.

Metadata That Should Accompany Every Artifact

Governance evidence becomes significantly more valuable when accompanied by standardized metadata.

Each governance artifact should, where appropriate, include:

- Document title
- Evidence domain
- Document owner
- Responsible department

- Date created
- Date approved
- Effective date
- Review frequency
- Last review date
- Current version
- Classification
- Related governance activity
- Retention category
- Repository location
- Approval authority

Standardized metadata improves searchability, authenticity, traceability, and long-term evidentiary value.

Catalog Maintenance

The Governance Evidence Catalog should not remain static.

Organizations should review it periodically to ensure it reflects:

- Changes in governance structure
- New regulatory requirements
- Emerging cyber risks
- Board reporting expectations
- Organizational growth
- Lessons learned from incidents
- Internal and external audit recommendations

As governance evolves, so should the catalog.

Using This Catalog

This catalog is intended to support—not replace—the Governance Readiness Scorecard and Evidence Gap Analysis.

A practical approach is to use it as follows:

- Inventory the governance evidence your organization currently maintains.
- Map each artifact to one of the five evidence domains.
- Identify missing, outdated, or poorly maintained evidence.
- Assign ownership for significant gaps.
- Incorporate improvements into the organization's governance roadmap.
- Review and update the catalog annually.

Organizations should resist the temptation to create every artifact listed in this appendix. The objective is not to accumulate documents. The objective is to ensure that meaningful governance activities consistently produce authentic, organized, and accessible evidence.

Ultimately, the value of this catalog lies not in the number of artifacts an organization possesses, but in its ability to demonstrate a clear, coherent, and defensible record of cybersecurity governance.

Appendix B—Board Cyber Governance Playbook

Purpose

Cybersecurity governance is a Board responsibility.

Cybersecurity operations are management's responsibility.

Confusing these two responsibilities is one of the most common governance mistakes organizations make. Boards should never manage firewalls, approve security tools, or direct incident response activities. Likewise, executive management should never expect the Board to perform operational cybersecurity functions.

The Board's role is to govern.

This playbook provides practical guidance for fulfilling that responsibility through the principles of the Defensible Evidence Framework™.

The Board's Governance Mission

The Board's objective is not to eliminate cyber risk.

It is to ensure that cyber risk is governed with reasonable care.

Every Board should strive to answer five questions confidently:

- Do we understand our organization's material cyber risks?
- Are those risks being managed appropriately?
- Are we receiving meaningful governance information?
- Are we exercising effective oversight?
- Can we demonstrate that oversight if challenged?

Everything in this playbook supports those five objectives.

The Board's Governance Responsibilities

Directors should focus on governance activities that cannot be delegated.

These include:

- Establishing governance expectations.
- Providing strategic oversight.
- Reviewing enterprise cyber risk.
- Approving governance policies.
- Confirming executive accountability.
- Monitoring management performance.
- Reviewing independent assurance.
- Evaluating governance maturity.
- Supporting continual improvement.

These responsibilities represent governance—not operations.

The Quarterly Board Agenda

Cybersecurity should appear on every regular Board agenda.

Quarterly discussions should generally include:

Governance

- Governance maturity updates
- Governance Readiness Score
- Significant governance changes
- Policy approvals
- Governance improvement initiatives

Risk

- Enterprise cyber risk profile
- Material risk changes
- Risk acceptance decisions
- Third-party risk
- Emerging threats

Oversight

- Executive cyber report
- Significant incidents
- Major projects
- Corrective action status
- Board requests from previous meetings

Assurance

- Internal audit updates
- External assessments
- Penetration testing summary
- Regulatory activity
- Governance validation

Executive Discussion

The Board should reserve time for discussion without presentations.

Governance happens during dialogue—not PowerPoint slides.

Questions Every Board Should Ask

Effective Directors ask governance questions.

Examples include:

Governance

- Has our governance structure changed?
- Is accountability still clear?
- Are governance responsibilities being fulfilled?
- Where is governance weakest today?

Risk

- What cyber risks concern management most?
- Which risks changed since our last meeting?
- What assumptions concern you?
- Which risks have we formally accepted?

Oversight

- What decisions require Board attention?
- Which corrective actions remain overdue?
- What keeps management awake at night?
- Where are we relying on assumptions?

Operations

- Have we exercised our incident response plan?
- What did we learn?
- Are critical vendors creating new risk?
- Are business continuity capabilities improving?

Assurance

- What independent assurance have we received?
- Which audit findings concern you most?
- Have previous findings been resolved?
- What evidence demonstrates improvement?

Questions Boards Should Avoid

Directors should resist becoming operational managers.

Poor governance questions include:

- Which firewall vendor are we using?
- Why wasn't server 14 patched?
- What SIEM rules were updated?
- Which antivirus product performed better?
- How many IDS alerts occurred yesterday?

These are management questions.

Instead ask:

- What governance decisions resulted from those findings?
- What enterprise risk do they create?
- Does management require Board support?

- What evidence demonstrates improvement?

The Board Cyber Dashboard

Every Board should receive a concise governance dashboard.

Rather than dozens of technical metrics, the dashboard should summarize information that supports fiduciary oversight.

Recommended sections include:

Governance

- Governance maturity level
- Governance Readiness Score
- Policy review status
- Governance initiatives

Enterprise Risk

- Top enterprise cyber risks
- Risk trend
- New material risks
- Accepted risks requiring Board awareness

Oversight

- Outstanding Board actions
- Open corrective actions
- Governance exceptions
- Executive reporting cadence

Operational Readiness

- Incident response readiness
- Tabletop exercise status
- Recovery testing status
- Third-party oversight

Assurance

- Internal audit findings
- External assessment summary
- Penetration testing summary
- Governance validation activities

One page is often enough.

Good dashboards support discussion rather than overwhelm Directors with data.

The Annual Governance Calendar

A mature Board establishes a predictable governance rhythm.

Quarter 1

- Annual cyber strategy review
- Governance Readiness Scorecard
- Enterprise risk review

Quarter 2

- Governance policy review

- Internal audit update
- Third-party governance review

Quarter 3

- Tabletop exercise participation
- Business continuity review
- Governance maturity discussion

Quarter 4

- Annual governance assessment
- Executive performance discussion
- Governance improvement planning

Consistency produces stronger governance.

Consistency also produces stronger evidence.

Board Participation During a Cyber Incident

The Board should not manage incident response.

Instead, Directors should focus on governance.

Board responsibilities include:

- Confirm executive leadership activation.
- Confirm incident governance procedures.
- Receive timely executive briefings.
- Understand business impact.
- Review legal and regulatory considerations.
- Monitor stakeholder communications.

- Support executive decision-making.
- Confirm after-action reviews occur.

The Board governs the organization while management manages the incident.

After the Incident

Following recovery, the Board should conduct a formal governance review.

Discussion should include:

- What governance worked well?
- What governance failed?
- Was oversight sufficient?
- Were decision authorities clear?
- Was the Board informed appropriately?
- What governance evidence was created?
- What improvements should be made?

The objective is governance improvement—not assigning blame.

Evidence the Board Should Expect

Directors should periodically verify that governance evidence exists for:

- Board cybersecurity minutes
- Committee minutes
- Governance dashboards
- Enterprise risk reports
- Risk acceptance decisions
- Governance policies

- Internal audit reports
- External assessments
- Tabletop exercise reports
- Corrective action tracking
- Governance Readiness assessments
- Evidence Gap Analyses

The Board should never assume these records exist.

It should periodically verify them.

Red Flags Every Board Should Recognize

The following conditions deserve immediate attention:

- Cybersecurity appears only after incidents.
- No formal cyber governance owner exists.
- Risk acceptance is undocumented.
- Governance evidence cannot be located quickly.
- Board reporting focuses only on technical metrics.
- Internal audit rarely evaluates cyber governance.
- Tabletop exercises exclude executive leadership.
- Corrective actions remain unresolved.
- Policies are outdated.
- Board education occurs infrequently.

One red flag may not indicate poor governance.

Several almost always do.

The Board's Annual Self-Assessment

At least annually, Directors should ask themselves:

- Do we understand our fiduciary responsibilities regarding cybersecurity?
- Are we receiving the right governance information?
- Do we challenge management appropriately?
- Do we understand the organization's most significant cyber risks?
- Can we explain our governance decisions?
- Would our governance withstand external scrutiny?
- Can our oversight be demonstrated through evidence?

The answers should guide the following year's governance priorities.

The Board Resolution

Every Board should periodically reaffirm its commitment to cybersecurity governance.

A simple statement captures the intent:

> **The Board of Directors recognizes cybersecurity as an enterprise governance responsibility. The Board commits to exercising reasonable oversight of cyber risk, maintaining clear accountability, supporting independent assurance, and ensuring that governance activities produce defensible evidence demonstrating responsible leadership.**

The Board's Legacy

Every Board eventually changes.

New Directors arrive.

Experienced Directors retire.

Chief executive officers move on.

Technology evolves.

Threats evolve.

The Board's governance record remains.

Years later, stakeholders will evaluate today's Board not by the number of cyber incidents that occurred, but by how leadership governed the organization before, during, and after those incidents.

That record becomes the Board's legacy.

The purpose of this playbook is to help ensure that legacy demonstrates thoughtful oversight, disciplined leadership, and governance that is not only effective—but defensible.

Appendix C—Governance Readiness Assessment

Purpose

The Governance Readiness Assessment provides organizations with a structured method for evaluating the maturity of their cybersecurity governance. Unlike technical security assessments that measure controls, vulnerabilities, or compliance, this assessment evaluates leadership's ability to demonstrate **reasonable cybersecurity governance through defensible evidence**.

The assessment aligns with the five evidence domains of the Defensible Evidence Framework™ and the Governance Maturity Ladder presented in this book.

The objective is not to achieve a perfect score.

The objective is to identify strengths, reveal evidence gaps, and establish priorities for continuous improvement.

Assessment Instructions

For each statement, select the score that best reflects your organization's current governance maturity.

Score	Description
1	Not Established
2	Initial / Informal
3	Defined and Repeatable
4	Well Managed and Measured
5	Fully Integrated and Defensible

Answer based upon **evidence**, not assumptions.

Whenever possible, verify the existence of supporting governance artifacts before assigning a score.

Domain 1 – Governance Evidence

Governance Structure

☐ Board oversight responsibilities are formally documented.
- ☐ Cybersecurity governance responsibilities are clearly assigned.
- ☐ Governance committees have current charters.
- ☐ Governance authority is formally delegated.
- ☐ Executive accountability is clearly documented.

Policy Governance

☐ Cybersecurity governance policies have Board approval.
- ☐ Policies are reviewed according to schedule.
- ☐ Policy revisions are documented.
- ☐ Policy exceptions follow a formal approval process.
- ☐ Governance calendars are maintained.

Governance Evidence Score: ______ / 50

Domain 2 — Risk Evidence

Enterprise Risk

- ☐ Enterprise cyber risks are documented.
    - ☐ Risk assessments occur on a regular schedule.
    - ☐ Business impacts are evaluated.
    - ☐ Material cyber risks are communicated to leadership.
    - ☐ Risk ownership is clearly assigned.

Risk Decisions

- ☐ Risk acceptance decisions are documented.
    - ☐ Executive approval is recorded.
    - ☐ Residual risk is evaluated.
    - ☐ Third-party cyber risks are governed.
    - ☐ Risk registers remain current.

Risk Evidence Score: ______ / 50

Domain 3 — Oversight Evidence

Board Oversight

- ☐ The Board receives regular cybersecurity reports.
    - ☐ Board minutes reflect meaningful cyber discussions.
    - ☐ Governance decisions are documented.
    - ☐ Corrective actions are monitored.
    - ☐ Directors receive cybersecurity governance education.

Executive Oversight

- ☐ Executive cyber governance meetings occur regularly.
    - ☐ Governance dashboards support leadership decisions.

- ☐ Escalation procedures are defined.
- ☐ Governance metrics are reviewed.
- ☐ Follow-up actions are tracked to completion.

Oversight Evidence Score: ______ / 50

Domain 4 — Operational Evidence

Preparedness

☐ Incident Response Plans are current.

- ☐ Tabletop exercises occur regularly.
- ☐ Business continuity plans are tested.
- ☐ Disaster recovery capabilities are validated.
- ☐ Crisis communications are documented.

Operational Governance

☐ Vendor oversight is documented.

- ☐ Security awareness is measured.
- ☐ Operational governance metrics are maintained.
- ☐ Lessons learned are documented.
- ☐ Corrective actions improve operations.

Operational Evidence Score: ______ / 50

Domain 5 — Assurance Evidence

Independent Assurance

☐ Internal Audit evaluates cyber governance.

- ☐ External assessments occur periodically.
- ☐ Penetration testing is performed.
- ☐ Corrective actions are independently verified.

☐ Governance maturity is periodically assessed.

Continuous Improvement

☐ Evidence Gap Analyses are conducted.
☐ Governance Readiness Assessments are repeated.
☐ Governance improvements are documented.
☐ Governance metrics demonstrate improvement.
☐ Independent assurance informs Board oversight.

Assurance Evidence Score: ______ / 50

Governance Evidence Repository

Evaluate the maturity of the organization's Governance Evidence Repository.

☐ Governance evidence is centrally organized.
☐ Evidence ownership is clearly assigned.
☐ Metadata standards are consistently applied.
☐ Version control is maintained.
☐ Governance evidence is searchable.
☐ Retention schedules are documented.
☐ Access controls protect sensitive evidence.
☐ Evidence can be produced quickly.
☐ Legal hold procedures exist.
☐ Repository reviews occur periodically.

Repository Score: ______ / 50

Overall Assessment

Appendix C—Governance Readiness Assessment

Domain	Maximum	Score
Governance Evidence	50	_____
Risk Evidence	50	_____
Oversight Evidence	50	_____
Operational Evidence	50	_____
Assurance Evidence	50	_____
Evidence Repository	50	_____

Total Score
_______ / 300

Governance Maturity Interpretation

Level 1 – Reactive (60–119)

Governance is largely informal and reactive. Leadership depends on individual knowledge rather than documented governance processes. Evidence is fragmented, incomplete, or difficult to locate.

Primary Objective: Establish governance structures and assign accountability.

Level 2 – Developing (120–179)

Governance processes are emerging. Policies, committees, and risk reporting exist but are inconsistent. Evidence quality varies across the organization.

Primary Objective: Standardize governance processes and improve evidence quality.

Level 3 – Managed (180–229)

Governance is repeatable and consistently executed. Leadership responsibilities are clear, evidence is organized, and oversight occurs regularly.

Primary Objective: Improve governance measurement and independent assurance.

Level 4 – Measured (230–269)

Governance effectiveness is actively measured. Evidence quality is monitored. Independent assurance validates governance activities, and leadership uses governance metrics to drive continual improvement.

Primary Objective: Integrate governance measurement across the enterprise and strengthen organizational culture.

Level 5 – Defensible (270–300)

Governance is fully integrated into organizational culture. Evidence is created naturally through disciplined leadership, readily accessible, independently validated, and capable of demonstrating reasonable cybersecurity governance under external scrutiny.

Primary Objective: Sustain governance excellence while continually adapting to evolving risks and business priorities.

Evidence Gap Summary

After completing the assessment, identify the organization's highest-priority improvement opportunities.

Governance Evidence

Risk Evidence

Oversight Evidence

Operational Evidence

Assurance Evidence

Governance Evidence Repository

Priority Improvement Plan

List the five most important governance improvements for the next twelve months.

Priority	Improvement	Owner	Target Date
1	____________	______	______
2	____________	______	______
3	____________	______	______
4	____________	______	______
5	____________	______	______

Executive Certification

Assessment Completed By:

Title: ------------------------------------

Date: ------------------------------------

Reviewed By: ------------------------------------

Board Committee (if applicable):

Final Reflection

The value of this assessment is not found in the score itself.

Its value lies in the conversations it creates.

Organizations that regularly assess their governance, honestly evaluate their evidence, and deliberately improve their leadership practices become progressively more resilient, more accountable, and more defensible.

The Governance Readiness Assessment should therefore be viewed not as a compliance exercise, but as an annual leadership discipline that demonstrates an organization's commitment to clarity, accountability, and defensible cybersecurity governance.

Appendix D—Sample Governance Evidence Repository Taxonomy

Purpose

A Governance Evidence Repository is more than a document library.

It is the institutional memory of cybersecurity governance.

When organized effectively, the repository enables leadership to quickly answer one of the most important questions that can arise after a cyber incident:

> **"Can you show us the evidence?"**

This appendix provides a sample taxonomy that organizations can adapt to fit their governance model. The taxonomy aligns with the five evidence domains of the Defensible Evidence Framework™ and is designed to make governance evidence searchable, traceable, and defensible.

The specific technology platform is less important than the organizational structure.

Whether the repository resides in SharePoint, Microsoft 365, a governance platform, a document management system, or another enterprise repository, the taxonomy should remain logical, consistent, and easy to navigate.

Repository Design Principles

Every Governance Evidence Repository should be:

- Organized around governance—not departments.
- Easy to navigate.
- Easy to search.
- Version controlled.
- Protected by appropriate access controls.
- Supported by standardized metadata.
- Maintained throughout the evidence lifecycle.

The repository should answer governance questions—not merely store files.

Recommended Top-Level Structure

```
Governance Evidence Repository
    ├── 01 Governance Evidence
    ├── 02 Risk Evidence
    ├── 03 Oversight Evidence
    ├── 04 Operational Evidence
    ├── 05 Assurance Evidence
    ├── 06 Governance Readiness
    ├── 07 Incident Governance
    ├── 08 Governance Metrics
    ├── 09 Historical Archives
    └── 10 Repository Administration
```

01 Governance Evidence

```
├── Board Governance
│   ├── Board Charters
```

```
|   ├──── Committee Charters
|   ├──── Board Resolutions
|   ├──── Governance Policies
|   └──── Governance Calendar
|
├──── Executive Governance
|   ├──── Executive Charters
|   ├──── Governance Committees
|   ├──── Delegations of Authority
|   └──── Accountability Matrix
|
├──── Policy Management
|   ├──── Approved Policies
|   ├──── Policy Reviews
|   ├──── Policy Exceptions
|   └──── Revision History
```

02 Risk Evidence

```
├──── Enterprise Risk Register
    ├──── Risk Assessments
    ├──── Business Impact Analysis
    ├──── Risk Acceptance
    ├──── Risk Treatment Plans
    ├──── Third-Party Risk
    ├──── Executive Risk Reports
    └──── Board Risk Reports
```

03 Oversight Evidence

```
├──── Board Minutes
    ├──── Committee Minutes
    ├──── Board Presentations
```

```
├──── Executive Dashboards
├──── Governance Reports
├──── Action Item Register
├──── Corrective Action Tracking
└──── Director Education
```

04 Operational Evidence

```
├──── Incident Response
│    ├──── Response Plans
│    ├──── Decision Logs
│    ├──── Situation Reports
│    └──── After-Action Reviews
│
├──── Tabletop Exercises
│
├──── Business Continuity
│
├──── Disaster Recovery
│
├──── Security Awareness
│
├──── Vendor Governance
│
└──── Operational Metrics
```

05 Assurance Evidence

```
├──── Internal Audit
├──── External Audit
├──── Penetration Testing
├──── Maturity Assessments
├──── Regulatory Reviews
```

```
    ├─── Governance Assessments
    ├─── Corrective Action Verification
    └─── Independent Reviews
```

06 Governance Readiness

```
├─── Governance Readiness Assessments
    ├─── Governance Readiness Scorecards
    ├─── Evidence Gap Analyses
    ├─── Governance Improvement Plans
    ├─── Annual Governance Reviews
    └─── Executive Action Plans
```

07 Incident Governance

This section should contain governance evidence created during major cyber incidents.

```
    ├─── Incident 2027-001
    │   ├─── Executive Decision Log
    │   ├─── Board Briefings
    │   ├─── Legal Reviews
    │   ├─── Regulatory Communications
    │   ├─── Insurance Documentation
    │   ├─── Governance Timeline
    │   ├─── Lessons Learned
    │   └─── Corrective Actions
    │
    ├─── Incident 2028-003
    │
    └─── Incident 2029-002
```

Each incident should become a complete governance case file.

08 Governance Metrics

├── Governance Dashboards
    ├── Quarterly Metrics
    ├── Annual Reports
    ├── KPI Trends
    ├── KGI Trends
    ├── Board Metrics
    └── Governance Benchmarking

09 Historical Archives

Historical governance evidence should never disappear.
    ├── Previous Policies
    ├── Previous Charters
    ├── Historical Risk Registers
    ├── Previous Assessments
    ├── Retired Dashboards
    ├── Historical Board Reports
    └── Archived Governance Programs

Historical evidence often becomes invaluable during litigation or regulatory review.

10 Repository Administration

├── Repository Standards
    ├── Metadata Standards
    ├── Naming Standards
    ├── Retention Schedule
    ├── Access Control Matrix
    ├── Repository Audit Log
    ├── Legal Hold Procedures

└──── Repository Owner Documentation

Recommended Metadata

Every governance artifact should include standardized metadata.

Metadata Field	Example
Document Title	Q2 2028 Board Cybersecurity Report
Evidence Domain	Oversight Evidence
Artifact Type	Board Presentation
Owner	Corporate Secretary
Business Unit	Enterprise Governance
Approval Authority	Board Risk Committee
Classification	Confidential
Created Date	2028-04-15
Approval Date	2028-04-30
Review Date	2029-04-30
Version	3.2
Retention Category	Board Records
Incident Reference	Incident 2028-003 (if applicable)
Related Documents	Risk Register 2028-Q2

Metadata should be mandatory wherever practical.

Recommended Naming Convention

A standardized naming convention dramatically improves searchability.

A recommended format is:

YYYY-MM-DD_Domain_DocumentType_Subject_Version

Examples:

2028-01-15_Governance_BoardMinutes_Q1_v1.0

2028-03-31_Risk_RiskRegister_Q1_v2.1

2028-04-22_Assurance_InternalAudit_CyberGovernance_v1.0

2028-07-12_Operations_Tabletop_RansomwareExercise_v1.2

2028-08-01_Oversight_BoardDashboard_Q3_v3.0

Avoid vague names such as:

- Board Meeting Final.docx
- Cyber Report NEW.pdf
- Policy Updated FINAL FINAL.docx

Those filenames create confusion over time.

Repository Security

Governance evidence frequently contains highly sensitive information.

Repositories should implement:

- Role-based access control
- Multi-factor authentication
- Encryption at rest
- Encryption in transit
- Immutable backups where appropriate
- Comprehensive audit logging
- Version history
- Legal hold capability
- Periodic access reviews

Protecting governance evidence is itself an exercise in governance.

Governance Ownership

Every repository should have clearly assigned ownership.

Typical responsibilities include:

Repository Owner

- Overall governance
- Repository standards
- Annual review

Corporate Secretary

- Board records
- Committee records

Enterprise Risk Management

- Risk evidence

Chief Information Security Officer

- Operational governance evidence

Internal Audit

- Assurance evidence

Legal

- Legal holds
- Retention requirements
- Ownership should be documented and reviewed annually.

Repository Health Checklist

Leadership should periodically ask:

- Can governance evidence be located within minutes?
- Are metadata standards consistently applied?
- Are outdated versions archived appropriately?
- Are retention schedules being followed?
- Are access permissions reviewed regularly?
- Are legal holds functioning properly?
- Does the repository reflect current governance?
- Would an external investigator understand our governance from this repository alone?

If the answer to any of these questions is "no," the repository requires attention.

Final Thought

A Governance Evidence Repository is not built for the day everything goes right.

It is built for the day everything goes wrong.

When a Board requests records.

When regulators issue inquiries.

When insurers ask for documentation.

When auditors begin their review.

When litigation begins.

Or when executive leadership simply wants to understand how governance has evolved over time.

A well-designed repository does more than organize documents.
It organizes the story of leadership.

And in the world of cybersecurity governance, that story may become one of the organization's most valuable assets.

Appendix F—Governance Evidence Crosswalk

Purpose

Organizations rarely operate under a single governance framework.

A healthcare organization may align with the Health Insurance Portability and Accountability Act (HIPAA), the National Institute of Standards and Technology (NIST) Cybersecurity Framework, and the Health Industry Cybersecurity Practices (HICP).

A financial institution may align with the Federal Financial Institutions Examination Council (FFIEC), NIST, and ISO/IEC 27001.

A defense contractor may align with NIST SP 800-171, CMMC, and contractual cybersecurity requirements.

Although these frameworks differ in structure and terminology, they share a common objective:

Demonstrating that cybersecurity is governed responsibly.

The Defensible Evidence Framework™ does not replace these standards.

Instead, it provides the evidentiary layer that demonstrates governance across them.

This appendix illustrates how the five evidence domains support widely recognized cybersecurity governance frameworks.

The Crosswalk Philosophy

Most cybersecurity frameworks answer one or more of the following questions:

- What should we protect?
- What risks should we manage?
- What controls should we implement?
- How should we respond?
- How should we recover?

The Defensible Evidence Framework™ asks a different question:

> **What evidence should exist if we are doing those things well?**

The relationship is complementary.

Frameworks define governance expectations.

The Defensible Evidence Framework™ demonstrates governance execution.

Crosswalk to NIST Cybersecurity Framework (CSF) 2.0

NIST CSF Function	Primary DEF™ Evidence Domain	Typical Governance Evidence
Govern	Governance Evidence	Board charters, governance policies, executive accountability, governance calendars
Identify	Risk Evidence	Enterprise risk register, business impact analyses, risk assessments
Protect	Operational Evidence	Security awareness records, policy implementation, operational procedures
Detect	Operational Evidence	Monitoring reports, detection governance, operational metrics
Respond	Operational Evidence	Incident response plans, decision logs, executive briefings
Recover	Operational Evidence	Disaster recovery testing, business continuity exercises, recovery metrics
Cross-Functional Validation	Assurance Evidence	Internal audits, maturity assessments, governance reviews

Crosswalk to NIST Cybersecurity Framework Govern Function

The Govern Function introduced in NIST CSF 2.0 aligns especially well with the Defensible Evidence Framework™.

NIST Govern Category	DEF™ Domain
Organizational Context	Governance Evidence
Risk Management Strategy	Risk Evidence
Roles, Responsibilities & Authorities	Governance Evidence
Policy	Governance Evidence
Oversight	Oversight Evidence
Supply Chain Governance	Risk Evidence / Operational Evidence
Governance Monitoring	Oversight Evidence
Continuous Improvement	Assurance Evidence

The Defensible Evidence Framework™ expands upon these governance concepts by identifying the evidence that demonstrates each governance activity occurred.

Crosswalk to ISO/IEC 27001

ISO/IEC 27001 Area	DEF™ Evidence Domain	Example Evidence
Leadership	Governance Evidence	Board approvals, governance policies
Planning	Risk Evidence	Risk treatment plans, enterprise risk register
Support	Operational Evidence	Awareness training, competence records
Operation	Operational Evidence	Incident response documentation, vendor governance
Performance Evaluation	Assurance Evidence	Internal audits, management reviews
Improvement	Assurance Evidence	Corrective actions, continual improvement records

Crosswalk to COBIT

COBIT Objective	DEF™ Evidence Domain
Governance System	Governance Evidence
Risk Optimization	Risk Evidence
Resource Optimization	Oversight Evidence
Performance Monitoring	Oversight Evidence
Delivery and Support	Operational Evidence
Assurance	Assurance Evidence

COBIT provides governance processes.

The Defensible Evidence Framework™ provides the evidence demonstrating those processes are functioning.

Crosswalk to COSO Enterprise Risk Management

COSO Component	DEF™ Evidence Domain
Governance & Culture	Governance Evidence
Strategy & Objective Setting	Risk Evidence
Performance	Operational Evidence
Review & Revision	Assurance Evidence
Information, Communication & Reporting	Oversight Evidence

Crosswalk to CMMC

CMMC Area	DEF™ Evidence Domain
Governance Documentation	Governance Evidence
Risk Management	Risk Evidence
Operational Security Practices	Operational Evidence
Assessment Readiness	Assurance Evidence
Continuous Monitoring	Oversight Evidence

Organizations pursuing CMMC certification often discover that governance evidence significantly simplifies assessment preparation.

Crosswalk to CIS Controls

CIS Controls	DEF™ Evidence Domain
Governance Policies	Governance Evidence
Risk Prioritization	Risk Evidence
Safeguard Implementation	Operational Evidence
Security Awareness	Operational Evidence
Assessment	Assurance Evidence
Management Reporting	Oversight Evidence

The CIS Controls describe operational safeguards.

The Defensible Evidence Framework™ demonstrates leadership oversight of those safeguards.

Crosswalk to Cyber Insurance

Cyber insurers increasingly evaluate governance—not merely technical controls.

Common underwriting questions align closely with the five evidence domains.

Insurance Question	DEF™ Domain
Does the Board oversee cybersecurity?	Governance Evidence
Are cyber risks evaluated regularly?	Risk Evidence
Does executive leadership monitor cyber risk?	Oversight Evidence
Are response capabilities tested?	Operational Evidence
Is governance independently assessed?	Assurance Evidence

Organizations with mature governance evidence often find it easier to support underwriting discussions and post-incident claims reviews.

Crosswalk to Regulatory Investigations

Following a significant cyber incident, regulators frequently request documentation demonstrating leadership oversight.

Typical requests map directly to the framework.

Regulatory Request	DEF™ Evidence Domain
Governance policies	Governance Evidence
Board oversight	Oversight Evidence
Risk assessments	Risk Evidence
Incident response documentation	Operational Evidence
Audit reports	Assurance Evidence

The framework therefore serves as a practical roadmap for preparing governance evidence before regulators request it.

Crosswalk to Board Fiduciary Responsibilities

Although terminology varies across jurisdictions, Boards generally exercise three core fiduciary responsibilities.

Fiduciary Responsibility	DEF™ Evidence Domain
Duty of Care	Oversight Evidence
Duty of Loyalty	Governance Evidence
Duty of Obedience / Compliance	Assurance Evidence
Enterprise Risk Oversight	Risk Evidence
Organizational Execution	Operational Evidence

The framework provides evidence supporting the Board's exercise of these governance responsibilities.

One Governance Activity, Many Frameworks

A single governance activity frequently satisfies multiple governance expectations simultaneously.

Consider an annual Board cybersecurity review.

It may produce:

- Board minutes
- Executive dashboard
- Updated enterprise risk register
- Policy approvals
- Risk acceptance decisions
- Corrective action assignments

That one meeting contributes evidence supporting:

- NIST CSF
- ISO/IEC 27001
- COBIT
- COSO ERM
- Cyber insurance underwriting
- Regulatory preparedness
- Board fiduciary oversight

This illustrates one of the framework's greatest strengths.

Organizations are not creating separate evidence for separate frameworks.

They are creating governance evidence once and using it many times.

The Value of the Crosswalk

One of the greatest misconceptions in cybersecurity governance is that every framework requires entirely different documentation.

In reality, mature governance produces evidence that supports numerous frameworks simultaneously.

The Defensible Evidence Framework™ organizes that evidence according to governance function rather than regulatory source.

This approach provides several advantages:

- Reduces duplicate documentation.
- Simplifies audit preparation.
- Improves regulatory readiness.
- Strengthens Board reporting.
- Supports cyber insurance reviews.
- Demonstrates governance maturity.

- Creates a single, coherent governance narrative.

Instead of asking, *"Which framework requires this document?"*, leadership begins asking a more valuable question:

> **"What governance evidence naturally results from responsible leadership?"**

That shift—from compliance-driven documentation to evidence-driven governance—is the defining principle of the Defensible Evidence Framework™.

Appendix E—Glossary

Assurance

Independent activities that evaluate whether governance, risk management, and operational practices are functioning as intended. Assurance commonly includes internal audits, external assessments, penetration tests, maturity assessments, and regulatory reviews.

Board Oversight

The Board of Directors' responsibility to oversee cybersecurity as an enterprise governance issue by establishing expectations, monitoring performance, reviewing material cyber risks, and holding executive leadership accountable.

Business Continuity

The capability of an organization to continue delivering critical business functions during and after a disruptive event through planning, preparedness, and recovery activities.

Corrective Action

A documented activity intended to eliminate an identified governance, security, or operational deficiency and prevent its recurrence.

Cybersecurity Governance

The system by which an organization's Board and executive leadership direct, oversee, and are accountable for managing cybersecurity risks in support of organizational objectives.

Decision Log

A chronological record of significant leadership decisions made during governance activities or cyber incidents, including participants, rationale, timing, and follow-up actions.

Defensible Evidence

Authentic, reliable, organized, and accessible documentation that demonstrates leadership exercised reasonable cybersecurity governance before, during, and after significant events.

Defensible Evidence Framework™ (DEF™)

A governance model that organizes cybersecurity governance evidence into five evidence domains—Governance, Risk, Oversight, Operational, and Assurance—to demonstrate reasonable cybersecurity governance.

Duty of Care

A fiduciary obligation requiring directors and executive leadership to make informed decisions with reasonable diligence, prudence, and oversight.

Enterprise Risk

A strategic risk capable of affecting an organization's objectives, financial performance, operations, reputation, legal obligations, or long-term sustainability.

Enterprise Risk Management (ERM)

A structured approach for identifying, evaluating, managing, and monitoring risks across the enterprise in support of organizational objectives.

Evidence

Information, documentation, records, or artifacts that demonstrate governance activities, decisions, oversight, accountability, or operational actions.

Evidence Domain

One of the five categories within the Defensible Evidence Framework™ used to organize governance evidence:

- Governance Evidence
- Risk Evidence
- Oversight Evidence
- Operational Evidence
- Assurance Evidence

Evidence Gap

The difference between the governance evidence that should reasonably exist and the evidence that is actually available.

Evidence Gap Analysis

A structured evaluation used to identify missing, incomplete, outdated, or inaccessible governance evidence and prioritize improvement efforts.

Evidence Lifecycle

The complete lifecycle of governance evidence, including creation, approval, use, maintenance, retention, and disposition.

Fiduciary Responsibility

The legal and ethical obligation of directors and executive leadership to act in the best interests of the organization through appropriate oversight, diligence, loyalty, and informed decision-making.

Governance

The processes, structures, leadership activities, and oversight mechanisms through which organizational objectives are directed and monitored.

Governance by Design

The practice of intentionally embedding governance responsibilities, accountability, decision-making, and evidence creation into organizational processes rather than treating governance as a reactive activity.

Governance Evidence

Evidence demonstrating that leadership established governance structures, assigned accountability, approved policies, delegated authority, and exercised organizational oversight.

Governance Evidence Catalog

A structured inventory of governance artifacts that organizations maintain to demonstrate cybersecurity governance maturity.

Governance Evidence Repository

A centralized, organized collection of governance artifacts that preserves the organization's cybersecurity governance history and supports rapid retrieval of evidence during audits, investigations, regulatory reviews, and cyber incidents.

Governance Maturity

The degree to which governance is consistently executed, measured, continuously improved, and supported by defensible evidence.

Governance Readiness

An organization's preparedness to demonstrate reasonable cybersecurity governance through credible evidence when examined by regulators, auditors, insurers, courts, or other stakeholders.

Governance Readiness Assessment

A structured evaluation of governance maturity across the five evidence domains of the Defensible Evidence Framework™.

Governance Readiness Scorecard

The scoring model used to measure governance maturity and readiness across Governance, Risk, Oversight, Operational, Assurance, and Governance Evidence Repository domains.

Governance Timeline

The chronological sequence of governance activities, leadership decisions, oversight actions, and supporting evidence surrounding a significant event.

Incident Governance

The leadership activities, oversight, decision-making, and documentation that occur while managing a cybersecurity incident.

Incident Response

The operational process of preparing for, detecting, analyzing, containing, eradicating, recovering from, and learning from cybersecurity incidents.

Institutional Memory

The accumulated knowledge, governance history, and documented decision-making preserved through organizational evidence.

Material Cyber Risk

A cybersecurity risk reasonably expected to influence executive decision-making, Board oversight, regulatory reporting, financial performance, or stakeholder confidence.

Metadata

Descriptive information associated with governance evidence that establishes authenticity, ownership, classification, dates, versions, approvals, and relationships to other governance activities.

Monday Morning Question

The central concept introduced in this book:

> **"If someone asked tomorrow to demonstrate reasonable cybersecurity governance, could your organization produce the evidence?"**

The Monday Morning Question serves as the practical test of governance readiness.

Operational Evidence

Evidence demonstrating that governance decisions have been implemented operationally through incident response, business continuity, security awareness, vendor oversight, recovery activities, and day-to-day security operations.

Oversight

Leadership activities that monitor governance effectiveness, evaluate management performance, review risk, and ensure accountability.

Oversight Evidence

Evidence demonstrating active Board and executive oversight through meeting minutes, governance dashboards, reports, corrective actions, and governance reviews.

Reasonable Cybersecurity Governance

Leadership oversight that reflects prudent, informed, documented, and accountable governance appropriate to the organization's size, complexity, industry, and risk environment.

Residual Risk

The level of risk that remains after security controls and mitigation activities have been implemented.

Retention Schedule

A documented policy specifying how long governance evidence should be preserved and when it may be archived or securely disposed of.

Risk Acceptance

A documented leadership decision acknowledging that a known level of residual cyber risk is acceptable based on organizational objectives and risk tolerance.

Risk Evidence

Evidence demonstrating how leadership identifies, evaluates, prioritizes, communicates, and governs cybersecurity risk.

Risk Register

A structured inventory of identified organizational risks, including likelihood, impact, ownership, mitigation activities, and current status.

Risk Tolerance

The amount and type of risk an organization is willing to accept in pursuit of its strategic objectives.

Tabletop Exercise

A facilitated discussion in which leadership and stakeholders walk through a simulated cyber incident to validate governance, decision-making, communications, escalation procedures, and organizational preparedness.

Third-Party Risk

Cybersecurity risks arising from vendors, suppliers, contractors, service providers, business partners, and other external organizations that access or support the enterprise.

Version Control

The disciplined management of document revisions to preserve governance history, maintain authenticity, and prevent unauthorized or accidental changes.

Governance Evidence Repository Taxonomy

The standardized organizational structure used to classify, store, search, and retrieve governance evidence throughout its lifecycle.

Final Thought

The terms defined in this glossary share a common purpose. They describe not only the language of cybersecurity governance, but also the concepts that enable organizations to demonstrate responsible leadership through evidence.

Throughout this book, one principle has remained constant:

Governance is exercised through leadership. Governance is demonstrated through evidence.

Appendix G—The Defensible Evidence Framework™ Quick Reference

Purpose

The Defensible Evidence Framework™ is designed to answer one fundamental question:

> **Can your organization demonstrate that it exercised reasonable cybersecurity governance?**

This Quick Reference summarizes the framework presented throughout this book. It is intended to serve as a practical reference for Boards, executive leadership, governance committees, auditors, legal counsel, risk professionals, and cybersecurity leaders.

Use it during Board meetings, governance reviews, tabletop exercises, governance assessments, and incident response activities.

The Core Principle

The strongest governance evidence is created naturally as governance occurs—not reconstructed after an incident.

Everything in the Defensible Evidence Framework™ builds upon this principle.

The Monday Morning Question

Every governance activity should ultimately prepare leadership to answer one question:

> **"If someone asked us tomorrow morning to demonstrate reasonable cybersecurity governance, could we produce the evidence?"**

If the answer is uncertain, governance requires improvement.

The Five Evidence Domains

1. Governance Evidence

Question Answered:

How did leadership establish cybersecurity governance?

Examples include:

- Governance policies
- Board charters
- Committee charters
- Delegations of authority
- Governance calendars
- Executive accountability
- Organizational structure

2. Risk Evidence

Question Answered:

How did leadership identify, evaluate, and govern cyber risk?

Examples include:

- Enterprise risk register
- Risk assessments
- Business impact analyses
- Risk acceptance decisions
- Third-party risk reviews
- Executive risk reports

3. Oversight Evidence

Question Answered:

How did leadership exercise oversight?

Examples include:

- Board minutes
- Committee minutes
- Board dashboards
- Executive governance reports
- Corrective action tracking
- Governance metrics

4. Operational Evidence

Question Answered:

How were governance decisions implemented?

Examples include:

- Incident response documentation
- Tabletop exercises
- Business continuity testing
- Disaster recovery testing
- Vendor oversight
- Security awareness records

5. Assurance Evidence

Question Answered:

How was governance independently validated?

Examples include:

- Internal audit
- External assessments
- Penetration testing
- Governance maturity assessments
- Corrective action verification
- Independent governance reviews

The Governance Evidence Flow

Every mature governance program follows the same progression.

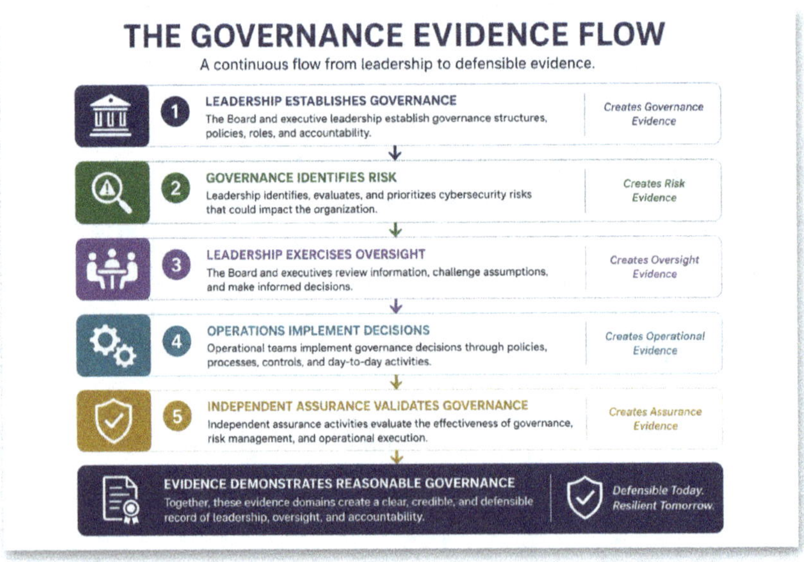

The framework is cumulative.

Each evidence domain strengthens the next.

Evidence Lifecycle

Governance evidence should progress through six lifecycle stages.

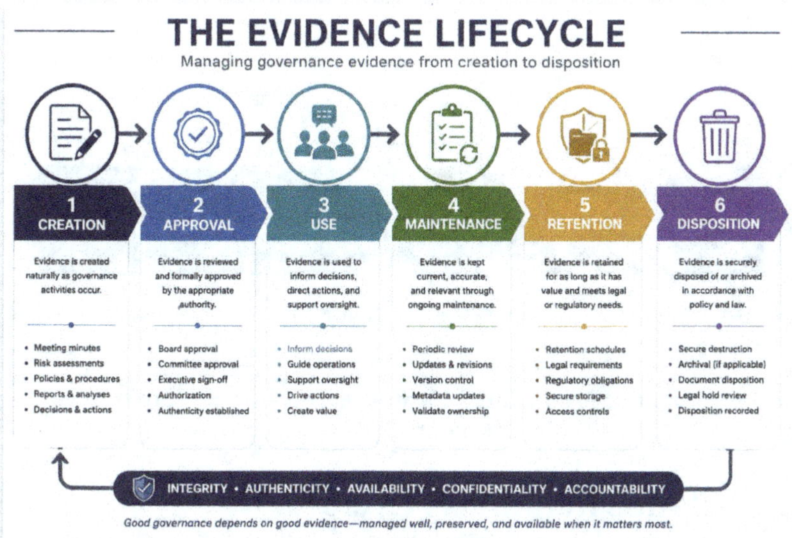

Every stage contributes to the credibility of the evidence.

Governance Maturity Ladder

Organizations progress through five governance maturity levels.

The objective is continual improvement—not perfection.

Governance Readiness Cycle

The framework is designed as a continuous improvement cycle.

Governance maturity is built through repetition.

Governance Before, During, and After an Incident

Before

- Establish governance
- Define accountability
- Build the repository
- Conduct tabletop exercises
- Measure readiness

- Perform Evidence Gap Analyses

During

- Govern the incident
- Maintain decision logs
- Preserve governance evidence
- Brief executive leadership
- Support Board oversight
- Document significant decisions

After

- Conduct governance review
- Perform after-action review
- Update policies
- Close corrective actions
- Improve governance
- Preserve incident evidence

The Governance Evidence Repository

Every organization should maintain a centralized Governance Evidence Repository.

The repository should be:

- Organized
- Searchable
- Version controlled
- Secure
- Metadata driven
- Regularly maintained

The repository preserves institutional memory.

Governance Health Checklist

Leadership should periodically ask:

- ☐ Are governance responsibilities clearly defined?
- ☐ Does the Board receive meaningful cybersecurity oversight?
- ☐ Are cyber risks formally documented?
- ☐ Are risk acceptance decisions recorded?
- ☐ Is governance evidence organized?
- ☐ Can governance evidence be located quickly?
- ☐ Are corrective actions tracked?
- ☐ Is governance independently assessed?
- ☐ Are Evidence Gap Analyses conducted?
- ☐ Is governance improving every year?

A "No" answer identifies an opportunity for improvement.

The Governance Conversation

Every Board meeting should answer four questions.

- What has changed?
- What risks require attention?
- What decisions require governance?
- What evidence demonstrates our oversight?

These questions keep governance focused on leadership rather than technology.

What Good Governance Produces

Well-governed organizations consistently produce:

- Better leadership decisions
- Clear accountability
- Stronger Board oversight
- Better executive reporting
- Higher-quality evidence
- Greater organizational resilience
- Increased stakeholder confidence
- Faster regulatory response
- Better audit outcomes
- Stronger cyber insurance positioning

Good governance creates value long before a cyber incident occurs.

Common Warning Signs

Organizations should investigate if they observe:

- Governance exists only on paper.
- Cybersecurity reaches the Board only after incidents.
- Risk acceptance is undocumented.
- Governance evidence is difficult to locate.
- Board reporting focuses exclusively on technical metrics.
- Corrective actions remain unresolved.
- Governance ownership is unclear.
- Internal Audit rarely evaluates cyber governance.
- Tabletop exercises exclude executives or Directors.
- Governance documentation is routinely recreated after requests.

These are symptoms of governance immaturity—not necessarily technical weakness.

The Framework in One Sentence

If the entire Defensible Evidence Framework™ could be summarized in a single sentence, it would be this:

> **Governance establishes accountability, accountability produces evidence, and evidence demonstrates that leadership exercised reasonable cybersecurity governance.**

The Final Reminder

Cybersecurity frameworks explain **what** organizations should do.

Security controls explain **how** organizations should protect systems.

The Defensible Evidence Framework™ explains **how leadership proves it governed responsibly.**

When governance is intentional, evidence becomes natural.

When evidence is natural, accountability becomes visible.

When accountability is visible, governance becomes defensible.

That is the enduring purpose of the **Defensible Evidence Framework™**.